Marika Vespa

Cement Uptake Mechanisms

Marika Vespa

Cement Uptake Mechanisms

Influence of the Inherent Heterogeneity of Cement on the Uptake Mechanisms of Ni and Co: A Micro-Spectroscopic Study

Südwestdeutscher Verlag für Hochschulschriften

Impressum / Imprint
Bibliografische Information der Deutschen Nationalbibliothek: Die Deutsche Nationalbibliothek verzeichnet diese Publikation in der Deutschen Nationalbibliografie; detaillierte bibliografische Daten sind im Internet über http://dnb.d-nb.de abrufbar.
Alle in diesem Buch genannten Marken und Produktnamen unterliegen warenzeichen-, marken- oder patentrechtlichem Schutz bzw. sind Warenzeichen oder eingetragene Warenzeichen der jeweiligen Inhaber. Die Wiedergabe von Marken, Produktnamen, Gebrauchsnamen, Handelsnamen, Warenbezeichnungen u.s.w. in diesem Werk berechtigt auch ohne besondere Kennzeichnung nicht zu der Annahme, dass solche Namen im Sinne der Warenzeichen- und Markenschutzgesetzgebung als frei zu betrachten wären und daher von jedermann benutzt werden dürften.

Bibliographic information published by the Deutsche Nationalbibliothek: The Deutsche Nationalbibliothek lists this publication in the Deutsche Nationalbibliografie; detailed bibliographic data are available in the Internet at http://dnb.d-nb.de.
Any brand names and product names mentioned in this book are subject to trademark, brand or patent protection and are trademarks or registered trademarks of their respective holders. The use of brand names, product names, common names, trade names, product descriptions etc. even without a particular marking in this works is in no way to be construed to mean that such names may be regarded as unrestricted in respect of trademark and brand protection legislation and could thus be used by anyone.

Coverbild / Cover image: www.ingimage.com

Verlag / Publisher:
Südwestdeutscher Verlag für Hochschulschriften
ist ein Imprint der / is a trademark of
AV Akademikerverlag GmbH & Co. KG
Heinrich-Böcking-Str. 6-8, 66121 Saarbrücken, Deutschland / Germany
Email: info@svh-verlag.de

Herstellung: siehe letzte Seite /
Printed at: see last page
ISBN: 978-3-8381-3502-1

Zugl. / Approved by: Zürich, ETH, Diss., 2006

Copyright © 2012 AV Akademikerverlag GmbH & Co. KG
Alle Rechte vorbehalten. / All rights reserved. Saarbrücken 2012

TABLE OF CONTENTS

Abstract 5

Zusammenfassung 7

Chapter 1: Introduction 9

    1.1 Objectives of this study 9

    1.2 Short introduction to cement 11

    1.3 Short introduction to the techniques 13

        1.3.1 Synchrotron-based techniques 13

            1.3.1.1 X-ray fluorescence 15

            1.3.1.2 X-ray absorption spectroscopy 16

        1.3.2 Scanning electron microscopy 17

    1.4 Outline of the thesis 18

    1.5 References 21

Chapter 2: Elemental, chemical and structural information in highly heterogeneous materials: A novel approach combing synchrotron-based X-rays and electron microscopic investigations 27

    Abstract 27

    2.1 Introduction 28

    2.2 Material and methods 30

        2.2.1 Sample preparation 30

        2.2.2 Scanning electron microscopy 31

        2.2.3 Synchrotron based-investigations 31

            2.2.3.1 Micro-XRF, Micro-XAS and micro-XRD data collection. 31

            2.2.3.2 Micro-XAS data reduction 33

    2.3 Results and discussion 33

        2.3.1 The role of cement in waste disposal 33

        2.3.2 Microscopic investigations 34

        2.3.3 Synchrotron-based investigations 37

            2.3.3.1 Micro-XRF investigations 37

            2.3.3.2 Ni speciation in Ni(II) enriched hydrated cement 39

            2.3.3.3 Co speciation in Co(II) enriched hydrated cement 42

            2.3.3.4 Oxidation of Cr in Cr(IV) enriched hydrated cement 44

            2.3.3.5 Al speciation in hydrated cement 46

  2.3.4 Synchrotron-based micro-XRD: An upcoming method for mineral phase identification in heterogeneous materials on the micro-level 48

  2.4 Conclusions.. 49

  2.5 References... 49

Chapter 3: Spectroscopic investigations of Ni speciation in hardened cement paste 57

  Abstract.. 57

  3.1 Introduction.. 58

  3.2 Materials and methods... 59

    3.2.1 Sample preparation... 59

    3.2.2 Wet chemistry experiments.. 60

    3.2.3 EXAFS data collection and reduction.. 60

    3.2.4 Diffuse reflectance spectroscopy (DRS).. 62

  3.3 Results and discussion... 62

    3.3.1 Wet chemistry data.. 62

    3.3.2 Formation of a Ni-Al LDH phase 64

    3.3.3 Time dependency of the Ni-Al LDH formation................................. 68

    3.3.4 Influence of other experimental parameters...................................... 69

    3.3.5 Controlling uptake mechanism of Ni in cement................................. 71

  3.4 References... 72

Chapter 4: Micro-scale investigations of Ni uptake by cement using a combination of scanning electron microscopy and synchrotron-based techniques 77

  Abstract.. 77

  4.1 Introduction.. 78

  4.2 Materials and methods... 79

    4.2.1 Sample preparation... 79

    4.2.2 Scanning electron microscopic investigations 79

    4.2.3 µ-XRF/µ-XAS data collection and reduction 80

  4.3 Results and discussion... 81

    4.3.1 Distribution and speciation of the Ni phases...................................... 81

    4.3.2 Influence of varying concentrations... 88

  4.4 References... 91

Supporting information to chapter 4 93
    S4.1 Details of µ-XAS data reduction 93
    Table S4.1 and Table S4.2 94
    Figure S4.1 95
    References 95

Chapter 5: The influence of hydration time on the Ni uptake by cement 97
    Abstract 97
    5.1 Introduction 98
    5.2 Material and methods 99
        5.2.1 Sample preparation 99
        5.2.2 µ-XRF and µ-EXAFS data collection and reduction 100
    5.3 Results and discussion 100
    5.4 References 106

Chapter 6: Co speciation in hardened cement paste: A macro- and micro-spectroscopic investigation 109
    Abstract 109
    6.1 Introduction 110
    6.2 Material and methods 111
        6.2.1 Sample preparation 111
        6.2.2 XAS and µ-XRF and data collection and reduction 112
    6.3 Results and discussion 113
        6.3.1 Influence of hydration time on the Co oxidation state 113
        6.3.2 Distribution and speciation of Co on the micro-scale 116
        6.3.3 The role of oxygen in Co(II) oxidation 119
        6.3.4 Implications of Co(II) oxidation in cementitious systems .. 120
    6.4 References 121

Supporting information to chapter 6 124
    Figure S6.1 124
    Table S6.1 125
    Fe K-edge measurements 126
    Electron paramagnetic resonance (EPR) 127
    Co(II)-sorbed onto HCP 128
    References 129

Chapter 7: Conclusions..131
    Outlook..135

Abstract

Cement-based materials are used worldwide for solidifying and stabilizing hazardous and radioactive wastes in order to prevent or retard the release of contaminants from the waste matrix into the environment. The long-term disposal of cement-stabilized hazardous waste is associated with landfilling of these waste forms, whereas deep geological disposal is foreseen for some categories of cement-stabilized radioactive waste. For the latter waste form, cement is used to condition and stabilize the waste materials and to construct the engineered barrier systems (container, backfill and liner materials).

The uptake mechanism of Ni(II) by hardened cement paste (HCP) at different reaction conditions (total metal concentration, hydration time, water/cement ratio, metal salts added to the system) were investigated by combining macro- and micro-spectroscopic with microscopic techniques. The macro- and micro-spectroscopic investigations (bulk- and µ-X-ray absorption spectroscopy, bulk- and µ-XAS) revealed for the Ni(II) speciation in HCP the predominant formation of Ni-Al layered double hydroxide (Ni-Al LDH) and minor amounts of Ni-hydroxide for all investigated systems. Furthermore, the bulk-XAS results showed that with ongoing hydration time the amount of Ni-Al LDH increased, whereas Ni-hydroxide decreased. This finding indicates that Ni-Al LDH is the thermodynamically stable phase in the long term, which likely controls the solubility of Ni in the Ni(II)-doped cement matrix. This finding was further supported by wet chemistry experiments. The only exception was observed at a low total Ni concentration of 50 mg/kg. In this system a Ni species predominantly forms, which could not be identified spectroscopically. Nevertheless, micro-scale investigations showed that Ni-Al LDH forms at highly enriched Ni regions. Scanning electron microscopy was employed to gain highly resolved spatial information on the element distribution and the interaction of Ni with cement phases in the Ni(II)-doped HCP system. The results reveal that for all of the investigated systems Ni forms rims around the non-hydrated mineral alite and its hydrated product inner-calcium-silicate-hydrate (inner-C-S-H). This finding indicates that the Ni-forming phases are directly related to specific cement minerals.

Macro- and micro-spectroscopic studies on Co(II)-doped HCP were performed to address the question whether the mechanistic model developed for the Ni immobilization in HCP can be applied to other transition metals. The results show, that Co is present in the oxidation states II and III in the cement matrix. In particular, the bulk-XAS results showed that the amount of Co(III) increases with hydration time, whereas Co(II) decreases. Co(II) is predominately incorporated into newly formed Co(II) hydroxide and/or Co-phyllosilicate, whereas Co(III) tends to be incorporated into a Co(III)OOH-like phase and/or Co-phyllomanganate.

Additional experiments were carried out to investigate the role of the oxidizing agents, in particular oxygen, on the Co speciation in the Co(II)-doped HCP samples. The result showed that Co is present in the oxidation state II in one sample, which was prepared under anoxic conditions. This indicates that oxygen has a strong influence on the Co oxidation state in cement. These findings suggest that Co(III)-containing phases and/or Co(III) surface-sorbed complexes should be taken into account for an overall assessment for Co-containing cement-stabilized waste under oxidizing conditions.

The findings from this study indicate that, although Ni(II) and Co(II) are both divalent cations, they react differently during the hydration of cement. Nevertheless, both cations became immobilized in specific minerals phases. These immobilization processes are expected to reduce the mobility of Co and Ni in the cement matrix and, consequently the impact on the environment. Furthermore, the study shows that immobilization processes in the cement matrix are specific with respect to the elements and cement phases that are involved.

Zusammenfassung

Für die Verfestigung und Stabilisierung von gefährlichen und radioaktiven Abfällen werden weltweit zementbasierende Materialen eingesetzt, damit die Freisetzung von gefährlichen Stoffen aus der Abfallmatrix in die Umwelt verhindert oder zumindest verzögert werden kann. Diese in Zement eingebundenen „Giftstoffe" werden dabei vielfach in Oberflächendeponien entsorgt. Für die sichere Endlagerung einiger Kategorien von zementverfestigten radioaktiven Abfällen sind vielfach geologische Tiefenlager vorgesehen. Dabei wird, neben der Herstellung von Abfallgebinden, Zement für die Konstruktion des Tunnelsystems und für die Verfüllung von Zwischenräumen verwendet.

Um das Verhalten von in Zement eingebundenen Metallen zu untersuchen, wurde in dieser Studie die Immobilisierung von Ni(II) und Co(II) in der Zementmatrix untersucht. Dabei wurden Untersuchungen mit herkömmlichen Mikroskopietechniken mit Makro- und Mikro-Spektroskopischen Untersuchungen kombiniert. Bei der Herstellung der mit Metall behandelten Proben wurden verschiedene Parameter variiert, wie z. B. die Metallkonzentration, die Hydratationszeit, das Wasser zu Zementverhältnis und die Metallsalze, welche zugefügt wurden. Die Makro- und Mikro-Spektroskopischen Untersuchungen mittels Röntgen-Absorptionsspektroskopie (XAS) ergaben für alle untersuchten Reaktionsbedingungen, dass die Einbindung von Ni(II) in Zement hauptsächlich durch die Neubildung von Ni-Al Doppelhydroxiden und nur zu kleinen Teilen durch die Ausfällung von Ni-Hydroxiden erfolgt. Es zeigte sich zudem, dass der Anteil der Ni-Al Doppelhydroxide mit der Dauer der Hydratationszeit zunahm, während die Anteile von Ni-Hydroxiden abnahm. Diese Resultate deuten daraufhin, dass die Ni-Al Doppelhydroxide im Ni-Zement-System die thermodynamisch stabilen Phasen sind und damit die Löslichkeit von Ni in diesen Systemen bestimmen können. In einem mit Ni(II) behandelten System mit der tiefsten Ni Konzentration (50 mg/kg), zeigte es ich auf der Makroskala (bulk-XAS), dass sich hauptsächlich eine Ni Spezies bildete, die bisher nicht identifiziert werden konnte. Die Untersuchungen auf der Mikroskala zeigten jedoch, dass sich auch in diesem System in Regionen, in denen Ni sehr stark angereichert war, sich Ni-Al Doppelhydroxide gebildet hatten.

Untersuchungen mittels Rasterelektronenmikroskopie wurden durchgeführt, um hochaufgelöste, ortsspezifische Informationen über die Elementverteilung und die Wechselwirkung von Ni mit Zementphasen im Ni-behandelten Zementsystemen zu gewinnen. Die Resultate aller untersuchten Systeme zeigten, dass Ni sich um Alite, beziehungsweise um dessen hydratisierte Form (innere-C-S-H) in Form eines Randes anreichert. Dies deutet darauf hin, dass die Neubildung von Ni-Phasen im hydratisierenden Zement in direkter Beziehung zur Bildung von bestimmten Mineralienphasen steht.

In der vorliegenden Arbeit wurde auch das Verhalten von Co(II) in Zement untersucht um die Frage zu beantworten, ob der für Ni(II) beobachtete Immobiliserungsprozess auch auf andere Übergangsmetalle zutrifft. Im Unterschied zu Ni(II) zeigten die XAS Messungen jedoch, dass nach anfänglicher Zugabe von Co(II) sich schon nach kurzer Hydratationszeit Co(III) in messbare Konzentrationen gebildet hatte. Die XAS Ergebnisse zeigten zudem, dass der Anteil von Co(III) mit zunehmender Hydratationszeit zunahm, während der Anteil von Co(II) abnahm. Die Ergebnisse zeigten im Weiteren, dass Co(II) hauptsächlich in Co(II)-Hydroxide und/oder in Co-Phyllosilikate eingebunden wird. Im Gegensatz dazu wird Co(III) bevorzugt in Co(III)OOH-ähnliche Phasen und/oder Co-Phyllomanganate eingebunden. Um den Einfluss von Luftsauerstoff auf den Oxidationsprozess nachweisen zu können, wurde auch eine Co(II) behandelte Probe in Abwesenheit von Sauerstoff präpariert. Die XAS Messungen zeigten, dass in dieser Probe Co(II) vorliegt. Dieses Ergebnis zeigt, dass der Sauerstoff einen starken Einfluss auf die Co Speziation in Zement ausübt. Aus diesen Ergebnissen darf geschlossen werden, dass unter oxidierenden Bedingungen in einer Zementmatrix Co(III) vorliegt, was in Beurteilungen bezüglich der Freisetzung dieses Metalls aus mit Zement stabilisierten Abfallmatrizen berücksichtigt werden sollte. Die Ergebnisse der vorliegenden Studie zeigen zweifelsfrei, dass Ni(II) und Co(II) sich bei der Einbindung in Zement unterschiedlich verhalten. Beide Metalle scheinen jedoch bevorzugt in neugebildeten Mineralstrukturen immobilisiert zu werden. Diese Mineralneubildungen können die Mobilität von Co(II)/Co(III) und Ni(II) in der Zementmatrix erheblich reduzieren. Die vorliegende Studie zeigt zudem, dass Immobilisierungsprozesse in der Zementmatrix sehr spezifisch erfolgen, beides hinsichtlich des involvierten Elements wie auch der beteiligten Zementphase.

CHAPTER 1

INTRODUCTION

1.1 Objectives of this study

Cement-based materials play an important role in multi-barrier concepts developed worldwide for the safe disposal of hazardous and radioactive wastes in order to prevent or retard the release of contaminants from the waste matrix into the environment. In Switzerland, the long-term disposal of cement-stabilized hazardous waste is associated with landfilling of these waste forms (e.g. 1), whereas deep geological disposal is foreseen for some categories of cement-stabilized radioactive waste (2). For example, more than 90 wt% of the near-field material of the planned Swiss repository for intermediate-level waste consists of hardened cement paste (HCP) and cementitious backfill materials. The cementitious near-field acts as engineered barrier for radionuclides and, consequently, prevents radionuclide migration into the host-rock. Hence, a detailed understanding of the chemical processes, which control the uptake of radionuclides by HCP, is a prerequisite for predicting the long-term behaviour in a cementitious repository. The immobilization potential of HCP originates from the selective binding properties for different chemical elements (e.g. 3), indicating that retention in cement systems is highly specific with respect to the mineral components and processes involved. From a chemical standpoint HCP is a very heterogeneous material with discrete particles typically in the size range of about a few hundred nanometers up to a few hundred micrometers. Overall, the material consists of mainly calcium silicate hydrates, portlandite: $Ca(OH)_2$, and calcium aluminates (for details see section 1.2). Furthermore, it contains ~9 wt% minor phases (e.g., hydrotalcite, hydrogarnet, ferrite etc.) that can act as highly reactive single mineral components (e.g. 4,5,6). For this reason the uptake of radionuclides by the complex cement matrix and individual cement phases such as C-S-H, portlandite etc. has been intensively studied in the past by combining wet chemical experiments with bulk X-ray absorption spectroscopic (bulk-XAS) experiments (e.g. 7,8-12).

XAS is an element specific non-invasive synchrotron-based technique which can provide atomic/molecular level information on the local coordination environment of an X-ray absorber. In many cases XAS investigations can yield a distinct molecular-level picture on the uptake processes of contaminants onto natural and man-made materials and allow to pin down the responsible mechanisms such as the chemical bonding of ions onto reactive binding sites of minerals and the formation of new phases (e.g. 13,14-16). In the past much of the understanding of surface processes has been obtained by resorting to bulk-XAS (macro-spectroscopy) measurements. In many cases such methodology is entirely sufficient to provide answers to pertinent questions, but the approach breaks down when mechanisms operative on the micro-scale have larger consequences. In the last few years, it has been realized that information on the micro-scale processes are needed if solutions for large scale problems such as toxic metal environmental contamination, soil remediation (e.g. 15,16-20) or nuclear waste storage (e.g. 21) have to be developed.

The main focus of the doctoral study is to investigate the influence of the inherent micro-scale spatial heterogeneity of cement on the immobilisation mechanisms of the metals Ni and Co.

Ni concentrations found in soils are normally <2 mmol/kg (22,23). Nevertheless, anthropogenic input (e.g., from tanneries, smelters or sewage sludge application, phosphate fertilizers, mining activities, auto and industrial emissions as well as municipal wastes) can increase the total nickel concentration in soils to >50 mmol/kg (23). Depending on the chemical speciation Ni(II) can become highly carcinogenic (e.g., NiO, $NiCO_3$, NiS). In radioactive waste isotopes of Ni (^{59}Ni $t_{1/2}$ = 7.5*10^4 a, ^{63}Ni $t_{1/2}$ = 100 a) are present mainly as long-lived activation products from metallic construction materials of nuclear power plants and other decommissioning waste. From previous studies it is known that Ni can form LDH-like phase in clay minerals and Al-oxides (e.g. 24,25,26) as well as in hydrated cement systems (4,5).

Co is released into the environment from burning coal and oil, from car and airplane exhausts, from municipal waste, and from a variety of industrial processes that use the metal or its compounds. Radioactive isotopes of Co, such as ^{60}Co ($t_{1/2}$ = 5.272 a) are used in nuclear medicine and in research. Co is an essential micronutrient required for the growth of plants and animals as well as a constituent of vitamin B12. Nevertheless, exposure to high Co concentrations through inhalation or ingestion over time can lead to severe health problems, such as heart function irregularities, high haemoglobin levels, lung cancer. The toxicity of cobalt depends on the solubility of the relevant compounds. In general, it is known that the less soluble form is also the less toxic (16).

The uptake of heavy metals by HCP and cement phases is not yet fully understood, in particular the distribution and speciation of heavy metals in the intact cement matrix. The present study was carried out with the aim of improving our understanding of Ni and Co uptake mechanism in cementitious systems. To achieve this goal sulphate-resisting Portland cement (CEM I 52.5 HTS, Lafarge, France), denoted as HTS (Haute Teneur en Silice), was treated with Ni and Co at selected reaction conditions (initial concentrations, water/cement ratios, anions present) and allowed to hydrate for different time periods. Crushed material and special thin sections were prepared from the hydrated HCP and investigated using various analytical methods.

The study is based on the use of µ-X-ray-absorption spectroscopy (XAS) with µ-X-ray fluorescence (µ-XRF) combined with scanning electron microscopy (SEM) to yield spatially-resolved molecular-level information on e.g., metal distribution, reactive cement phases, redox processes, surface complexation, formation of new mineral phases occurring in the Ni/Co-doped HCP system. The µ-XRF/XAS and SEM studies were complemented with bulk-XAS to verify the overall relevance of the prevailing mechanism observed by the micro-spectroscopic studies.

To summarise, the main research questions of this study were the following:
- Are Ni and Co immobilized within the cement matrix, and if yes, under which form?
- If more than one Ni and Co phase form, which is the prevailing mineral phase?
- Is there a relationship between the formations of newly-formed Ni and Co phases with cement minerals?
- Is there an influence of the experimental reaction conditions (hydration time, initial concentrations, water/cement ratios, anions present), on the Ni and Co uptake mechanism?

1.2 Short introduction to cement

Cement is mainly produced from limestone and additives such as, clays, iron ore or bauxite. The raw material is heated up to a temperature of ~1500°C. The temperature is regulated so that the product results in sinter, but not molten mass or glass. In a first heating step, low temperatures are reached, during which the calcium carbonate turns to calcium oxide and calcium dioxide. In a second step, higher temperatures are reached. Throughout this process the calcium oxides and silicates react to form tricalcium and dicalcium silicates (alite: Ca_3SiO_5; belite: Ca_2SiO_4). At the same time, small amounts of tricalcium aluminates ($3CaO \cdot Al_2O_3$) and tetracalcium aluminoferrite (ferrite, $4CaO \cdot Al_2O_3 \cdot Fe_2O$) are also formed. The resulting material is called 'clinker' and the phases formed are referred to as clinker phases. To achieve the desired quality, gypsum (which controls the setting and hardening time) can be added to the finished product. At this stage the material is pulverized and ready for use.

The cement used in this study belongs to the so called 'Portland cement', which, per definition, consists mainly of alite and belite. Portland cement is the most common type of cement in general usage and it is the basic ingredient of concrete and mortar. Five different types of Portland cement with varying compositions exist. Sulphate-resisting Portland cement denoted as HTS cement was used in this study. In a sulphate-resisting Portland cements it is important to lower the content of tricalcium aluminates, which, if in contact with sulphate-rich fluids after initial setting and hardening, may react to form ettringite and/or monosulphates. This process opens up cracks, which create new connected porosity accelerating the transport of sulphate into the cement paste, and consequently a faster deterioration of the material.

The clinker material used in this study contains ~61 wt% alite, 18 wt% belite, 5.3 wt% ferrite, 3.9 wt% tricalcium aluminates, 3.7 wt% calcium carbonate, 3.6 wt% calcium sulphates and <2wt% minor oxides (27).

Soon after the clinker material is mixed with water an intense reaction occurs referred to as 'hydration'. During this process the clinker phases dissolve to create the so-called hydrated phases to give the hardened cement paste (HCP) its properties, which depend mainly on the water/cement (w/c) ratio. In this hydration process alite is one of the first phases to react followed by belite. Alite and belite decompose to form calcium silicate hydrates (C-S-H, $(CaO)_x(SiO_2)_y(H_2O)_z$ with CaO/SiO_2 between 0.8-1.8, $SiO_2/H_2O < 0.5$) and calcium hydroxide (portlandite, ~18 wt%). C-S-H is the most abundant hydrated cement phase with ~46 wt% and confers most of its properties to the HCP. Tricalcium aluminates are the fastest reactive phases together with the calcium and alkali sulphates. Note that the dissolution of the alkali sulphates confers the high alkalinity to the cement (pH>13). The ferrites have an intermediate reaction time between alite and belite. Both aluminates and ferrites form the aluminoferrite monosulphate (AFm: $Ca_2(Al,Fe)(OH)_6)\cdot X\cdot H_2O$) and aluminoferrite trisulphate (AFt, e.g. ettringite $Ca_6Al_2(SO_4)_3(OH)_{12}\cdot 26H_2O$) hydrated minerals (~17 wt%). Further minor phases, such as hydrotalcite ($Mg_{1-x}Al_x(OH)_2]^{x+}(A^{n-})_{x/n}\cdot yH_2O$, ~1.5 wt%) are also formed during hydration. Further ~15.6 wt% clinker phases and minor amounts of oxides and calcium carbonate are still present after the hydration process is 'completed'. The hydration process is assumed to be completed after ~30 days. After this time ~90% of the cement is hydrated. Nevertheless, the reactions continue to further hydrate the cement at much slower rate.

1.3 Short introduction to the techniques
1.3.1 Synchrotron-based techniques

Since the 1950's particle accelerators have been developed and were originally designed and mainly used for high-energy physics experiments, but shortly after retrofitted with X-ray beamlines. Soon it was recognized that the X-rays generated from particles accelerators could be exploited for a wider range of scientific fields, such as in material research, geo-and environmental science as well as in biology. Therefore, the X-ray research community promoted the realization of modern synchrotron sources with dedicated X-ray beamlines. Advances in X-ray optics as well as the extremely high brilliance of modern synchrotron radiation sources has led to today's use of focused X-ray beam sizes from micrometer to nanometer or less (e.g., 28). The spatial resolution of state of the art beamlines is ~1x1 μm^2 using KB mirrors (Kirkpatrick-Baez) (29), whereas the use of zone plates allows to achieve resolution of ~30 nm (e.g., scanning transmission X-ray microscopic beamlines (STXM (30,31)). One must bear in mind that, in addition to the x-y resolution, the penetration depth z has to be taken in account. For example, for an energy of 10 keV in a SiO_2 matrix the penetration depth amounts to ~200 μm, whereas at 4 keV in the same matrix, the penetration depth only amounts to ~15 μm (calculation performed by using the computer program HEPHAESTUS (32)). Examples of synchrotron-based micro-focused X-ray techniques are μ-X-ray fluorescence (μ-XRF) and μ-X-ray absorption spectroscopy (μ-XAS). μ-XRF combined with μ-XAS allows spatially resolved in-situ investigations to be performed, the local chemical environment and the structure of the element of interest to be discerned without any prior knowledge of the long-range crystallographic order.

X-ray beamlines which serve for μ-XRF and μ-XAS investigations can be roughly classified into two categories: tender and hard X-ray beamlines with energies typically <4 keV and >4 keV, respectively. Tender X-ray beamlines are appropriate to investigate mainly K-edges of light elements, i.e. elements with atomic number between C and Ca in the periodic system. On the other hand, hard X-ray beamlines are the method of choice for studying the chemical environment of heavier elements. In this case, the K- or L-edges of elements with atomic numbers above Ca in the periodic system are probed. In cases, where both light (Z<~20) and heavier elements (Z>~20) are involved in the process under investigation, it is beneficial to combine the information of both tender and hard X-rays to gain a more complete and detailed system understanding. Specimens for XAS measurements in the tender X-rays energy region have to be placed in vacuum or He environment due to the high absorption of the X-rays in air. The XAS measurements in the hard X-ray energy region, however, can be performed in air.

The experimental setup is relative simple. The XAS measurements can be performed in transmission and fluorescence modes. Measurements in transmission mode work well when the concentration of the X-ray absorber in the sample exceeds about 2 wt%. For transmission mode measurements the sample is positioned between two detectors, e.g. two ion chambers (Fig. 1.1).

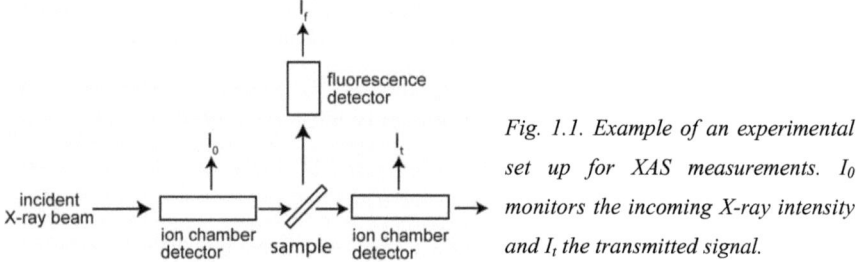

Fig. 1.1. Example of an experimental set up for XAS measurements. I_0 monitors the incoming X-ray intensity and I_t the transmitted signal.

The first chamber (I_0) measures the incoming intensity, whereas the second (I_t) measures the intensity of the transmitted beam.

For fluorescence mode measurements, employed for both XRF and XAS, the sample is usually positioned at an angle of 45° between the first chamber and the fluorescence detector, which is often a solid state detector (I_f) (Fig. 1.1). The fluorescence mode is the method of choice for samples without a transmitted signal, such as thick compact samples, and for dilute samples. For the latter, the measurements must be based on the core-hole created by photoelectron emission which is produced by the impinging X-rays (Fig. 1.2a). In the following, the relaxation process consists of either fluorescence (Fig. 1.2b) and/or Auger electron emission (Fig. 1.2c). Auger and fluorescence emission are inversely related; with increasing atomic number electron production decreases and X-ray production increases. Because of the short penetration depth of electrons the detection of Auger electrons is useful in surface studies. The disadvantage of Auger electron measurements is the requirement for vacuum conditions. For natural samples which usually contain water this requirement is problematic, but one can overcome this problem by measurements in bulk sensitive fluorescence mode. Detectors used for fluorescence measurements are Stern-Heald-type detectors (Lytle detector, The EXAFS Co.) and energy discriminating solid-state detectors.

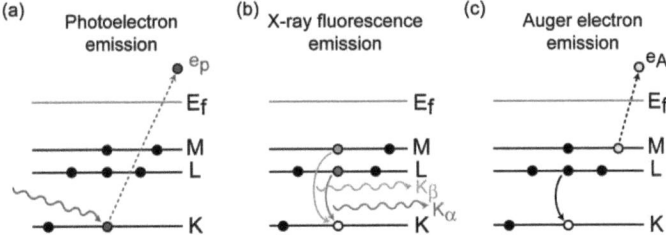

Fig. 1.2. Schematic illustration of a) core-hole production caused by impinging X-rays and ejection of the core electron (e_p) into the continuum (above the Fermi energy, E_f); b) filling of the vacancy produced by the photoelectric effect by electrons form higher energy shells (e.g., L or M; radiative decay) resulting in a release of energy ($K\beta/K\alpha$), c) filling of the vacancy produced by the photoelectric effect by electrons from higher energy shells resulting in the emission of Auger electrons (e_A, nonradiative decay).

The detection limit for XAS measurements in fluorescence mode depends very much on the matrix of the sample material. For example, XAS spectra of a heavy metal in solution with concentration in the range of a few mg/kg can easily be detected. In contrast, the investigation of 50 mg/kg Ni in a heterogeneous cementitious matrix, which contains up to a few wt% Fe, can take up to ~15 hours of EXAFS data collection in order to gain a sufficient good signal/noise ratio (33).

1.3.1.1 X-ray fluorescence

X-ray fluorescence (XRF) is a powerful tool to obtain spatially-resolved chemical elemental distribution overview maps (e.g., 2x2 mm^2) from heterogeneous matrices. The information gained from such maps is a qualitative elemental distribution relative to the concentration of the elements present in the investigated matrix. The μ-XRF maps are an important first step for localizing the areas of interest for μ-XAS investigations.

The μ-XRF maps are obtained by scanning the sample under the monochromatic beam at a given energy. Below this energy (e.g., 10 keV) the core electrons of certain chemical elements are excited (e.g., Ca-Zn K-edges) and the fluorescence emission signal of the exited elements is detected. The advantage of being able to tune the beam energy is that it enables mapping the sample below and above a characteristic absorption edge of a chemical element. This approach allows enhancing the chemical contrast by calculating difference maps, collected below and above the absorption edge. Furthermore, the position and the fine structure of an absorption edge reveal information on the chemical speciation of the absorber. For example, depending on the oxidation state of an absorber there is a shift in the position of the absorption edge. This energy shift can be used to gain chemical contrast by mapping the region of interest at the characteristic energies of the different oxidations states (34).

1.3.1.2. X-ray absorption spectroscopy

The potential of the XAS technique lies in the determination of the chemical speciation of unknown species within a complex heterogeneous matrix. These species can be crystalline, amorphous, surface-sorbed or soluble. XAS provides structural and chemical information on the crystal-chemical environment within a distance of ~5 Å surrounding the X-ray absorber. The information gained by XAS includes the coordination environment, bond distances between atoms, type of nearest neighbours, degree of disorder and oxidation state.

When the X-ray energy is equal or slightly larger than the binding energy of the X-ray absorber, a core electron is emitted by photoelectric processes and a sharp increase in the absorption spectrum is observed (Fig. 1.3).

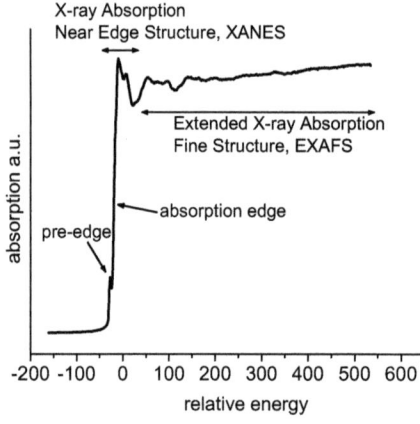

Fig. 1.3. Schematic illustration of an X-ray absorption spectroscopic spectrum indicating the pre-edge, the absorption edge, XANES and EXAFS regions explained in the text.

XAS can be divided into two regions: X-ray absorption near edge structure (XANES) and extended X-ray absorption fine structure (EXAFS) (35) (Fig. 1.3). The XANES region is dominated by electron transitions and multiple-scattering events near the absorption edge of the absorbing atom, which covers the energy range between ~20 eV below and ~50 eV above the absorption edge. XANES is mainly used to extract information on the oxidation state, based on the edge position, and for fingerprinting by comparing experimental spectra with reference compounds. Some elements also exhibit a so-called pre-edge feature (Fig. 1.3), which is due to interatomic transitions into free bound states below the Fermi energy, and which can also be used for fingerprinting and to extract information on the oxidation state. EXAFS, on the other hand, is the region extending from ~50 eV up to ~2000 eV above the absorption edge. In the EXAFS region multiple scattering can be neglected in a first approximation. In this case, the photoelectron can be regarded as a photoelectron wave emanating from the absorbing atom, which is elastically scattered due to the interaction with the inner electrons of the neighbouring atoms. Depending on the phase of the waves (in-going and out-going) this interaction results in a constructive or destructive interference and a slight modification of the absorption coefficient of the absorbing atom. Thus, the interference caused by the absorption coefficient in the EXAFS region shows a series of oscillations which may extend up to 2000 eV above the absorption edge (35-38).

1.3.2 Scanning electron microscopy

Scanning electron microscopy (SEM) is a suitable tool to retrieve high quality micrographs, and to conduct semi-quantitative chemical analysis. SEM has been used for several decades in a large number of scientific areas e.g., geology, geochemistry, crystallography, biology, etc. (e.g., 39,40,41). Modern SEM microscopes have a high spatial resolution typically in the range of a few hundred nanometres. SEM microscopes are usually equipped with secondary electron (SE) and backscatter electron (BSE) detectors, for macrographs, and an energy dispersive spectrometer (EDS) for analytical purposes. The SE are generated at the surface of the sample, within the first nanometre (Fig. 1.4), caused by the interaction of the electron beam with the core electrons of the atoms. The SE images can be obtained on unpolished surfaces and provide information on the topographic contrast of the sample at high resolution. In contrast, BSE are generated within the first micrometer of the sample. BSE images create a specific phase contrast thanks to which, minerals can be identified according to their brightness in the image. The minerals with the greatest average atomic number show the brightest contrast, whereas those with the lower atomic number show darker contrast (42,43). The combination of SE, BSE with EDS detectors allows complementary information on the microstructure and the chemical elemental distribution to be obtained. Furthermore, through EDS-single spot analysis, the elemental composition of single minerals can be determined. It should be noted that electron beams can cause sample damage and, thus, samples should be handled carefully under the beam, in particular for EDS-point analysis. For example, hydrated phases may disappear, as in the case of calcium silicate hydrates (C-S-H) (44). To minimize damage, the electron beam is usually defocused, the time of the sample to be exposed under the electron beam is minimized, or if implemented in the microscope, the sample stage is cooled. The high vacuum (10^{-6} mbar) normally used in common SEM can also cause sample damaging. Thus, some microscopes have been adjusted for environmental studies, consenting, e.g., hydrated phases (e.g., C-S-H) to be investigated without causing artefacts due to water loss. These environmental scanning electron microscopes (ESEM) have the advantage of allowing the adjustment of the vacuum between 2-40 mbar.

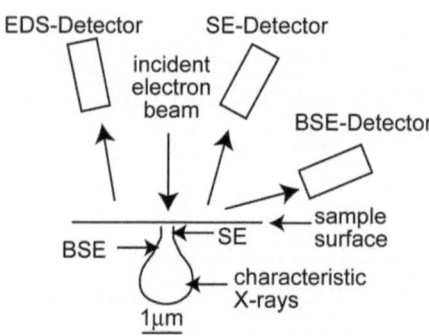

Fig. 1.4 Scanning electron microscopic experimental set up. The bulb indicates the penetration of the electrons in the sample through the incident beam. Through the interaction of the electron beam with the sample secondary electrons (SE), backscatter electrons (BSE) and characteristic X-rays are generated. The SE and BSE signals are detected from a photomultiplier (SE-and BSE-detector), whereas characteristic X-rays signals are collected from a Si(Li) detector (EDS-detector).

1.4 Outline of the thesis

The main part of the Ph.D. thesis is composed of five chapters, which correspond to papers subject to different stages of the publication process.

Chapter 2:

Elemental, chemical and structural information in highly heterogeneous materials: A novel approach combing synchrotron-based X-rays and electron microscopic investigations. M. Vespa, R. Dähn, D. Grolimund, M. Harfouche, E. Wieland, A. M. Scheidegger. To be submitted to Journal of Microscopy.

The present chapter illustrates a powerful multi-analytical approach by the synergetic use of scanning electron microscopy, synchrotron-based µ-X-ray fluorescence, µ-X-ray spectroscopy and µ-X-ray diffraction. The study illustrates how such a multi-analytical approach can be used to gain specially-resolved information on the elemental distribution, chemical speciation and morphology of complex and heterogeneous systems. This information can be obtained from the same sample and on the same area of interest on the micro-scale. All the wealth of this information is essential to develop a better molecular understanding of heavy metals immobilization by hydrated cement paste.

Chapter 3:

Spectroscopic investigations of Ni speciation in hardened cement paste (2006). M. Vespa, R. Dähn, D. Grolimund, E. Wieland, A. M. Scheidegger. Environmental Science and Technology, 40, 2275-2285.

In this chapter, X-ray absorption spectroscopy (XAS) and diffuse reflectance spectroscopy (DRS) were used to determine the local environment of Ni in hardened cement paste using crushed Ni(II)-doped materials.. The bulk-XAS study showed that for all investigated systems Ni(II) is predominantly immobilized in a layered double hydroxide (LDH) phase. This finding was supported by DRS measurements. Only a small amount of Ni(II) precipitates as Ni-hydroxides (α-Ni(OH)$_2$ and β-Ni(OH)$_2$), indicating that Ni-Al LDH rather than Ni-hydroxides are the thermodynamically most stable form in cementitious systems. Thus, Ni-Al LDH may control the solubility of Ni(II) in Ni-doped hardened cement paste.

Chapter 4:

Micro-scale investigations of Ni uptake by cement using a combination of scanning electron microscopy and synchrotron-based techniques.
M. Vespa, R. Dähn, E. Gallucci, D. Grolimund, E. Wieland, A. M. Scheidegger. Environmental Science and Technology, in press.

In this chapter, information on the cement microstructure, Ni distribution, Ni concentration and speciation of the Ni phases formed in the cement system and their association with specific cement minerals are presented. Scanning electron microscopic investigations show that for all metal loadings (50 up to 5000 mg/kg) the Ni phases form rims around inner-calcium silicate hydrates, suggesting a direct association with this cement phase. μ-X-ray absorption spectroscopic (μ-XAS) measurements further revealed that a mixture of Ni-Al LDH and only a small amount of Ni-hydroxides (α-Ni(OH)$_2$ and β-Ni(OH)$_2$) form in Ni enriched regions. A comparison of the results from the micro-spectroscopic investigation with those from the earlier macro-spectroscopic study (Chapter 3) indicates that the same Ni phases form both on the macro- and micro-scale. At 50 mg/kg Ni loading, however, the μ-

XAS measurements suggested the presence of an additional Ni species. In the latter system Ni-Al LDH is found in Ni-rich regions, whereas the unknown species forms in Ni-poor regions.

Chapter 5:
The influence of hydration time on the Ni uptake by cement.
M. Vespa, R. Dähn, E. Wieland, D. Grolimund, A. M. Scheidegger. Czechoslovak Journal of Physics, in press.

This chapter presents the results of the influence of the hydration time on the Ni speciation in hardened cement paste. Information on the Ni distribution and speciation of the Ni phases formed in the cement system has been gained by employing µ-XRF and µ-XAS. The Ni-doped samples investigated in this study were hydrated for six hours and one year to account for the chemical environment in a fresh and aged cement paste, respectively. The µ-XAS measurements revealed that a mixture of Ni phases form at single regions of interests, independently of the hydration time. Data analysis further supported the findings from the previous studies that Ni(II) is predominantly immobilized in a Ni-Al LDH and that only small amounts of Ni-hydroxide form. The important finding in this study, however, was that the portion of Ni-Al LDH increases with the hydration time, whereas the amount of Ni-hydroxides decreases. This further supports the idea that Ni-Al LDH is the thermodynamically most stable Ni species in the cement matrix.

Chapter 6:
Co speciation in hardened cement paste: A macro- and micro-spectroscopic investigation.
M. Vespa, R. Dähn, D. Grolimund, E. Wieland, A. M. Scheidegger. Submitted to Environmental Science and Technology.

X-ray-absorption spectroscopy (XAS) has been used to determine the local environment of Co in crushed Co(II)-doped cement samples. Further, in-situ investigations on the micro-scale were carried out using µ-XAS and µ-XRF. µ-XRF was used to gain information on the Co distribution, whereas µ-XAS was employed to determine the speciation and oxidation state of Co on the micro-scale. The Co(II)-doped cement samples were prepared under normal atmosphere, to simulate similar conditions as for the waste packages. To investigate the role of oxygen a further sample was prepared under conditions in which oxygen content was minimized. The unhydrated cement powder was mixed by adding a

Co(NO$_3$)$_2$ solution at a water/cement ratio of 0.4. Subsequently, the samples were hydrated for different time periods from 1 hour up to 1 year. The study showed that for the samples prepared in air Co(II) is oxidized to Co(III) after 1 hour of hydration time. Moreover, the amount of Co(III) increases with increasing hydration time, whereas Co(II) decreases. The μ-XRF revealed the high heterogeneity of Co in the cement matrix in respect to distribution and speciation. The study also revealed that Co forms spot-like and ring-like structures. μ-XAS showed that Co(II) is predominately incorporated into newly formed Co-hydroxide-like phase and/or Co-phyllosilicates, whereas Co(III) tends to be incorporated into CoOOH-like phase and/or Co-phyllomanganates. On the contrary, Co was found to be present as Co(II) in the sample prepared under conditions of reduced oxygen exposure. This finding indicates that oxygen has a strong influence on the Co speciation in hardened cement paste. The findings further suggest that Co(III) species or Co(III)-containing phases should be taken into account for an overall assessment of the Co release from Co-containing cement-stabilized waste under oxidizing conditions.

1.5 References

(1) Schmidt, M.; Beckefeld, P.; Götz, R.; Kamsties, S.; Kretz, C.; Molitor, N.; Neck, U.; Vogel, P. *Reststoff-und Abfallverfestigung. Immobilisierung von Schadstoffen-Recycling-Verbesserung der Deponiefähigkeit*; Expert Verlag: Renningen-Malmheim, **1995**.

(2) Chapman, N.; McCombie, C. *Principles and standards for the disposal of long-lived radioactive wastes*; First ed.; Elsevier Science, Ltd.: Oxford, **2003**.

(3) Glasser, F. P. Chemistry of cement-solidified waste forms. . In: *Chemistry and microstructure of solidified waste forms*; Spence, R. D., Ed.; Lewis Publishers: Boca Raton, **1993**.

(4) Scheidegger, A. M.; Wieland, E.; Dähn, R.; Spieler, P. Spectroscopic evidence for the formation of layered Ni-Al double hydroxides in cement. *Environmental Science and Technology* **2000**, *34*, 4545-4548.

(5) Scheidegger, A. M.; Wieland, E.; Scheinost, A. C.; Dähn, R.; Tits, J.; Spieler, P. Ni phases formed in cement and cement systems under highly alkaline conditions: An XAFS study. *Journal of Synchrotron Radiation* **2001**, *8*, 916-918.

(6) Rose, J.; Bénard, A.; J., S.; Borschneck, D.; Hazemann, J.-L.; Cheylan, P.; Vichot, A.; Bottero, J.-Y. First insights of Cr speciation in leached portland cement using X-ray spectromicroscopy. *Environmental Science and Technology* **2003**, *37*, 4864-4870.

(7) Bonhoure, I.; Scheidegger, A. M.; Wieland, E.; Dähn, R. Iodine species uptake by cement and CSH studied by K-edge X-ray absorption spectroscopy. *Radiochimica Acta* **2002**, *90*, 647-651.

(8) Bonhoure, I.; Wieland, E.; Scheidegger, A. M.; Ochs, M.; Kunz, D. EXAFS study of Sn (IV) immobilization by hardened cement paste and calcium silicate hydrates. *Environmental Science and Technology* **2003**, *37*, 2184-2191.

(9) Kirkpatrick, R. J.; Brown, G. E.; Xu, N.; Cong, X. X-ray absorption spectroscopy of C-S-H and some model compounds. *Advances in Cement Research* **1997**, *9* (33), 31-36.

(10) Allen, P. G.; Siemering, G. S.; Shuh, D. K.; Bucher, J. J.; Edelstein, N. M.; Langton, C. A. Technetium speciation in cement waste forms determined by X-ray absorption fine structure spectroscopy. *Radiochimica Acta* **1997**, *76*, 77-86.

(11) Richard, N.; Lequex, N.; Boch, P. An X-ray absorption study of phases formed in high-alumina cements. *Advances in Cement Research* **1995**, *7* (28), 159-169.

(12) Ziegler, F.; Scheidegger, A. M.; Johnson, C. A.; Dähn, R.; Wieland, E. Sorption mechanisms of zinc to calcium silicate hydrate: X-ray absorption fine structure (XAFS) investigation. *Environmental Science and Technology* **2001**, *35*, 1550-1555.

(13) Fendorf, S. Fundamental aspects and applications of X-ray absorption spcetroscopy in clay and soil science. In: *Synchrotron X-ray methods in clay science*; Schulze, D. G., Stucki, J. W., Bertsch, P. M., Eds.; The Clay Mineral Society, **1999**.

(14) Allen, P. G.; Bucher, J. J.; Denecke, M. A.; Kaltsoyannis, N.; Nitsche, H.; Reich, T.; Shuh, D. K. Applications of synchrotron radiation techniques to chemical issues and environmental sciences. In: *Synchrotron radiation techniques in industrial, chemical, and material science* Amico, D., Ed.; Plenum press: New York, **1996**.

(15) Manceau, A.; Lanson, B.; Schlegel, M.; Hargé, J. C.; Musso, M.; Eybert-Bérard, L.; Hazemann, J.-L.; Chateigner, D.; Lamble, G. Quantitative Zn speciation in smelter-contaminated soils by EXAFS spectroscopy. *American Journal of Science* **2000**, *300*, 289-343.

(16) Manceau, A.; Marcus, M.; Tamura, N. Quantitative speciation of heavy metals in soils and sediments by Synchrotron X-ray techniques. In: *Application of synchrotron radiation in low-temperature geochemistry and environmental science*; Fenter, P. A., Rivers, M. L., Sturchio, N. C., Sutton, S. R., Eds.; Mineralogical Society of America: Washington, DC, **2002**.

(17) Manceau, A.; Marcus, M. A.; Tamura, N.; Proux, O.; Geoffroy, N.; Lanson, B. Natural speciation of Zn at the micrometer scale in a clayey soil using X-ray fluorescence, absorption, and diffraction. *Geochimica et Cosmochimica* **2004**, *68* (11), 2467-2483.

(18) Manceau, A.; Tommaseo, C.; Rihs, S.; Geoffrey, N.; Chateigner, D.; Schlegel, M.; Tisserand, D.; Marcus, M.; Tamura, N.; Chen, Z.-S. Natural speciation of Mn, Ni, and Zn at the micrometer scale in a clayey paddy soil using X-ray fluorescence, absorption, and diffraction. *Geochimica et Cosmochimica Acta* **2005**, *69* (16), 4007-4034.

(19) Catalano, J. G.; Heald, S. M.; Zachara, J. M.; Brown Jr., G. E. Spectroscopic and diffraction study of uranium speciation in contaminated Vadose Zone sediments from the Hanford Site, Washington State. *Environmental Science and Technology* **2004**, *38*, 2822-2828.

(20) Nachtegaal, M.; Marcus, M.; Sonke, J. E.; Vangronsveld, J.; Livi, K. J. T.; Van Der Lelie, D.; Sparks, D. L. Effects of *in situ* remediation on the speciation and bioavailability of zinc in a smelter contaminated soil. *Geochimica et Cosmochimica Acta* **2005**, *69* (19), 4649-4664.

(21) Zachara, J. M.; Ainsworth, C. C.; Brown, G. E. J.; Catalano, J. G.; McKinley, J. P.; Qafoku, O.; Smith, S. C.; Szecsody, J. E.; Traina, S. J.; Warner, J. A. Chromium speciation and mobility in high level nuclear waste Vadose zone plume. *Geochimica et Cosmochimica Acta* **2004**, *68*, 13-20.

(22) Adriano, D. C. *Trace elements in the terrestrial environment*; Springer, **1986**.

(23) Uren, N. C. Forms, reactions, and availability of nickel in soils. *Advances in Agronomy* **1994**, *48*, 141-203.

(24) Scheinost, A. C.; Sparks, D. L. Formation of layered single- and double-metal hydroxide precipitates at the mineral/water interface: A multiple-scattering XAFS analysis. *Journal of Colloid and Interface Science* **2000**, *223*, 1-12.

(25) Scheidegger, A. M.; Lamble, G. M.; Sparks, D. L. Spectroscopic evidence for the formation of mixed-cation hydroxide phase upon metal sorption on clays and aluminum oxides. *Journal of Colloid and Interface Science* **1997**, *168*, 118-128.

(26) Scheidegger, A. M.; Strawn, D. G.; Lamble, G. M.; Sparks, D. L. The kinetics of mixed Ni-Al hydroxide formation on clay and aluminum oxide minerals: A time-resolved XAFS study. *Geochimica et Cosmochimica Acta* **1998**, *62* (13), 2233-2245.

(27) Lothenbach, B.; Wieland, E. A thermodynamic approach to the hydration of sulphate-resisting portland cement. *Waste Management* **2006**, *26*, 706-719.

(28) Schulze, D. G.; Stucki, J. W.; Bertsch, P. M. *Synchrotron X-ray methods in clay science*; The Clay Minerals Society, **1999**.

(29) Yang, B. X.; Rivers, M. L.; Schildkamp, W.; Eng, P. GeoCARS micro-focusing Kirkpatrick-Baez mirror bender development. *Reviews of Scientific Instruments* **1995**, *66*, 2278-2280.

(30) Tyliszczak, T.; Warwick, T.; Kilcoyne, A. L. D.; Fakra, S.; Shuh, D. K.; Yoon, T. H.; Brown, G. E., Jr.; Andrews, S.; Chembroulu, V.; Strachan, J.; Acremann, Y. Soft X-ray scanning transmission microscope working in an extended energy range at the Advanced Light Source. *Synchrotron Radiation Instrumentation, AIP Conference Proceedings* **2003**, *705*, 458-461.

(31) Warwick, T.; Ade, H.; Fakra, S.; Gilles, M.; Hitchcock, A.; Kilcoyne, A. L. D.; Shuh, D. K.; Tyliszczak, T. Further development of soft X-ray scanning microscopy with an elliptical undulator at the Advanced Light Source. *Synchrotron Radiation News* **2003**, *16* (3), 22-27.

(32) Ravel, B.; Newville, M. ATHENA, ARTEMIS, HEPHAESTUS: data analysis for X-ray absorption spectroscopy using IFEFFIT. *Journal of Synchrotron Radiation* **2005**, *12*, 537-541.

(33) Vespa, M.; Dähn, R.; Grolimund, D.; Wieland, E.; Scheidegger, A. M. Spectroscopic investigation of Ni speciation in hardened cement paste. *Environmental Science and Technology* **2006**, *40*, 2275-2282.

(34) Grolimund, D.; Senn, M.; Trottmann, M.; Janousch, M.; Bonhoure, I.; Scheidegger, A. M.; Marcus, M. Shedding new light on historical metal samples using micro-focused

synchrotron X-ray fluorescence and spectroscopy. *Spectrochimica Acta Part B* **2004**, *59*, 1627-1635.

(35) Koningsberger, D. C.; Prins, R. *X-ray absorption*; John Wiley & Sons: New York, **1987**.

(36) Brown Jr., G. E. Spectroscopic studies of chemisorption reaction mechanisms at oxide-water interfaces. In: *Mineral-Water Interface Geochemistry*; Hochella, M. F., White, A. F., Eds.: Washington, DC., **1990**.

(37) Charlet, L.; Manceau, A. Structure, formation, and reactivity of hydrous oxide particles: Insights from X-ray absorption spectroscopy. In: *Environmental particles*; Buffle, P. J., van Leeuwen, H. P., Eds.; Lewis Publishers: Boca Raton, **1993**.

(38) Fenter, P. A.; Rivers, M. L.; Sturchio, N. C.; Sutton, S. R. *Applications of synchrotron radiation in low-temperature geochemistry and environmental science*; The Mineralogical Society of America: Washington DC, **2002**.

(39) Buseck, P. R.; Self, P. Electron energy-loss spectroscopy (EELS) and electron channeling (ALCHEMI). In: *Minerals and reactions at the atomic scale: Transmission electron spectroscopy*; Buseck, P. R., Ed.; Mineralogical Society of America: Blacksburg, Virginia, **1992**.

(40) Potts, J. P.; Bowles, J. F. W.; Reed, S. J. B.; Cave, M. R. *Microprobe techniques in the earth sciences*; First ed.; Chapman & Hall: London, **1995**.

(41) Hayat, M. A. *Principles and technique of electron microscopy biological applications*; Fourth ed.; Cambridge University Press: Cambridge, **2000**.

(42) Scrivener, K. L. Backscatter electron imaging of cementitious microstructures: understanding and quantification. *Cement and Concrete Composites* **2004**, *26*, 935-945.

(43) Vespa, M.; Dähn, R.; Gallucci, E.; Grolimund, D.; Wieland, E.; Scheidegger, A. M. Micro-scale investigation of Ni uptake by cement using a combination of scanning electron microscopy and synchrotron-based techniques. *Environmental Science and Technology* **2006**, in press.

(44) Taylor, H. F. W. *Cement Chemistry*; Second ed.; Thomas Telford: London, **1997**.

CHAPTER 2

ELEMENTAL, CHEMICAL AND STRUCUTRAL INFORMATION IN HIGLY HETEROGENEOUS MATERIALS: A NOVEL APPROACH COMBING SYNCHROTRON-BASED X-RAYS AND ELECTRON MICROSCOPIC INVESTIGATIONS

Abstract

A powerful analytical approach is presented to investigate physiochemical processes in complex heterogeneous materials. The approach is based on the combined use of scanning electron microscopy and synchrotron-based micro X-ray fluorescence (XRF), micro X-ray absorption spectroscopy (XAS) and micro X-ray diffraction (XRD). In the present study we discuss how this approach can help to gain spatially-resolved information on the element distribution, the chemical speciation and the mineral phases observed. Complementary elemental, chemical and structural information can be obtained with micro-scale resolution on the same sample and the same areas of interest.

The multi-technique approach described in this paper is well suited to discern the correlation among chemical elements present in the matrix and the coordination environment of selected elements of interest. Furthermore, it allows information on the microstructure and mineral phases present in complex and highly heterogeneous materials to be gained. The potential of the complementary multi-technique approach will be outlined by presenting an example of heavy metals immobilization by solidified hydrated cement. Cement is a very heterogeneous material with discrete mineral particles in the nano- to micrometer size range and is, therefore, well-suited to illustrate the novel multi-technique analytical approach.

2.1 Introduction

Spatially-resolved elemental, chemical and structural information is often the crucial key to decipher physiochemical processes in highly heterogeneous materials. For example, soils and cement are materials which consist of an inherent mixture of different mineral phases. The most reactive fractions of these mineral assemblies have particle sizes in the nanometer to micrometer range. Moreover, the metal speciation in such complex systems may vary spatially over a few hundred micrometers. Highly heterogeneous materials may be analyzed in a first step by a variety of analytical techniques such as laser-ablation induced coupled mass spectrometry (LA-ICP-MS), transmission electron microscopy (TEM) and scanning electron microscopy (SEM). These are well established methods for acquiring average compositional, chemical and morphological information. For example, SEM microscopes are equipped with a back scattering electron (BSE) detector and energy dispersive (EDS) micro-analysis. BSE imaging is based on the spatial variation of the electron density and allows the optical identification of typical mineral phases present in mineral assemblages based on the grey level contrast and the morphology (e.g., 1,2-4). Spatially resolved information on the microstructure and phase association is obtained from BSE imaging due to the high magnification that can be achieved. The combination of BSE imaging with EDS micro-analysis, further, allows spatially resolved semi-quantitative information on the chemical composition of the different mineral phases to be gained (e.g., 5). While LA-ICP-MS, TEM, SEM-based BSE and EDS are essential techniques to further improve our understanding of the microstructure, phase association and the chemical composition of mineral phases in complex matrices with spatial resolution, no information on the micro-scale chemical speciation on elements of interest can be obtained.

Only few analytical techniques are suitable for micro-scale studies on the speciation of elements in heterogeneous materials. Synchrotron based FTIR-spectro-microscopy and laser Raman microanalysis (LRMA) provide molecular information with a spatial resolution below 10 µm. However, the sensitivity of the above techniques is limited, and the interpretation of the resulting spectra is not straightforward. Electron beam techniques, on the other hand, have the best spatial resolution (nanometer range) that is presently achievable with analytical techniques (thereby approaching physical limits). Electron energy loss spectroscopy (EELS) allows micro-speciation information to be gained by analyzing the fine-structure of the element specific edges in the energy-loss spectrum (6). EELS appears as an excellent tool for specialized questions requiring chemical information on a sub-micrometer-scale, e.g. analysis of the colloidal fraction (7). Nevertheless, the laborious sample preparation and the occurrence of possible radiation damages can be a problem (8).

The most promising technique for investigating the chemical speciation of elements in heterogeneous materials is synchrotron-based X-ray absorption spectroscopy (XAS). The method allows the speciation of chemical entities in complex natural and engineered materials to be determined, even when the element of interest is present at low concentrations (concentration of X-ray absorber down to a few tens of ppm). Unknown species can be identified in crystalline as well amorphous materials, or when the X-ray absorber is present as a surface-sorbed species or even in solution. XAS is a local probing technique which provides molecular-level information of an X-ray absorbing atom of interest within a distance of ~5 Å. Most frequently used XAS techniques are: X-ray absorption near edge structure (XANES) and extended X-ray absorption fine structure (EXAFS) (e.g., 9). XANES is mainly used to discern the oxidation state of the X-ray absorber, based on the edge position, and for fingerprinting by comparing experimental spectra of unknown species with reference compounds. EXAFS, on the other hand, is employed to gain information on the coordination sphere (i.e., type of neighboring atoms, bond length and coordination numbers) of the X-ray absorber (9-12).

In general, XAS does not yield spatially-resolved structural data since the dimension of the X-ray beam is much bigger (at most beamlines >100 x 100 μm^2) than the typical particle size of minerals (normally < 20 x 20 μm^2). Therefore, in the past, much of our understanding of chemical speciation of elements of interest, in complex natural and engineered materials, has been obtained by resorting to bulk-XAS measurements on powder materials (13-20). In bulk-XAS measurements the chemical speciation of an X-ray absorber of interest is determined indirectly from the averaged XAS signal generated from all individual species. In many cases, such methodology is entirely sufficient to provide answers to pertinent questions regarding the coordination environment of the element of interest in complex matrices, but the approach breaks down when mechanisms operative on the micro-scale have larger consequences. In the last few years, it has been realized that solutions for large scale problems, such as toxic metal environmental contamination as well as remediation and nuclear waste storage, should be based on detailed information on micro-scale processes. In view of the importance of small-scale processes and molecular-level mechanisms in complex heterogeneous systems, there has been a considerable effort to develop high resolution analytical synchrotron-based X-ray probes with which the wealth of structural information provided by XAS can be obtained on a micro-scale (21-31). For in-depth reviews on micro-probe beamlines and applications of micro-spectroscopy the reader is referred to Bertsch and Hunter (32), Manceau et al. (33) and Sutton et al. (34).

A key advantage of synchrotron-based X-ray micro-beam analytical facilities, is the fact that they combine high photon flux, high brilliance and high wavelength tuneability with micro-beam opportunities and high spatial stability. Micro-X-ray fluorescence (XRF), micro-XAS and micro-X-ray diffraction (XRD) experiments can be combined to obtain elemental, structural and crystallographic information in complex heterogeneous materials. In such combined micro-beam experiments micro-XRF is essential, in a first stage, to map the partitioning of trace contaminants among coexisting mineral phases in the investigated sample. Micro-XAS opens up the possibility to identify the different mechanisms of metal uptake on a molecular level. Finally, micro-XRD allows the determination of small particles and is, thus, suited for micron size phase identifications. However, one must keep in mind that the quality of XRD is degraded as crystallite size decreases. Therefore, the success of micro-XRD experiments declines as grain size decreases to sub-micron dimensions.

Within this manuscript we present examples on metal immobilization processes occurring in solidified hydrated cement. The examples are given with the aim of illustrating the potential of the combined use of SEM-based BSE images and EDS micro-analyses with synchrotron-based micro-XRF, micro-XAS and micro-XRD. The multi-technique approach is demonstrated on the same regions of interest in given samples. The examples focus on identifying the speciation of metals added to the cement matrix (e.g, Ni, Co, Cr) as well as on the speciation of elements inherent to the cement system (e.g, Al). Cement is chemically a very complex and heterogeneous material with particle in the nano- to micrometer size range, and is, therefore, appropriate to illustrate this multi-technique approach.

2.2 Materials and methods
2.2.1 Sample preparation

The cement samples were prepared from a commercial sulphate-resisting Portland cement (CEM I 52.5 N HTS, Lafarge, France). Metal-enriched hydrated cement pastes were prepared by mixing a $Ni(NO_3)_2$, $Co(NO_3)_2$ or K_2CrO_4 solution with unhydrated cement. The metal salts were dissolved in deionized water to obtain stock solutions with concentrations of 0.3 mol/L (pH=~3-6). The solutions were mixed with the unhydrated cement at a water/cement (w/c) ratio of 0.4 according to the European Norm EN-196-3. The final metal concentrations of Ni(II), Co(II) and Cr(VI) in the pastes were 5000 mg/kg. The cement pastes were filled into Plexiglas moulds, which were closed with a polyethylene lid, and hydrated for

30 days. After a hydration time of 30 days the cement is to 99.9% compacted and hardened. The samples were stored in closed containers at 100% relative humidity. The cylinders were cut into several slices of ~1 cm thickness and dried in the glovebox (dry N_2 atmosphere, CO_2 and O_2 <2 ppm, T=20±3 °C). Some slices were impregnated and polished for the preparation of thin sections. The polished thin sections, prepared at the Spectrum Petrographics, Inc. (USA), were supported with "Instant Grazy Glue" (Elmer's Products Inc.). The "Instant Grazy Glue" allows the thin section to be removed from the supporting glass slide. This allows micro-XRD measurements in transmission mode to be performed without the background contribution from the supporting glass slide. The thin sections were employed for both SEM-based BSE/EDS investigations and synchrotron-based micro-focused XRF/XAS/XRD measurements.

2.2.2 Scanning electron microscopy

The SEM investigations were conducted at the Laboratory for Construction Materials (IMX), Ecole Polytechnique Fédéral de Lausanne (EPFL) using a FEI Quanta 200 microscope and at the Laboratory for Materials Behaviour (LWV) at the Paul Scherrer Institute (PSI) using a Zeiss DSM 962 microscope. The FEI microscope was operated at an accelerating voltage of 15 kV and a beam current of 100 µA, whereas the Zeiss microscope was operated at an accelerating voltage of 20 kV and a beam current of 76 µA. The FEI microscope is equipped with a solid state detector, whereas the Zeiss is equipped with a photomultiplier for BSE imaging. The difference in the grey scale resulting on the BSE images, which allows identifying different mineral phases, can be further used to filter these mineral phases by the use of computer programs, such as NIH-image and, thus, create false colour maps as shown in Fig. 2.1.

Both microscopes are further equipped with a Si(Li)detector for EDS analysis. The sample-volume probed was $\sim 1\mu m^3$.

2.2.3 Synchrotron-based investigations
2.2.3.1 Micro-XRF, micro-XAS and micro-XRD data collection

Micro-XRF maps reflect the fluorescence signal detected from the sample under investigations. The information gained from such maps is a qualitative elemental distribution. The beamline on which the data are collected is crucial. Every beamline has an optimized energy range that depends among others on the X-ray source (e.g., bending magnet, insertion device), the monochromator crystals and the X-ray absorber materials in the beam path (e.g., X-ray windows). Furthermore, micro-beam facilities offer specific experimental set ups, allowing experiments to be conducted in air or in vacuum.

In this study, micro-XRF maps of Ca, Si, Al and S were performed on the LUCIA beamline at the Swiss Light Source (SLS), Switzerland (29). The micro-XRF maps were obtained by scanning the sample in a tender vacuum under the monochromatic beam with a beam size of 10x10 µm^2. For Ca, Si and Al the micro-XRF maps were recorded at the energy of 4.1 keV. In the case of S mapping, the beam energy was set to 3.9 keV (below the Ca K-edge) in order to avoid the saturation of the detector by the Ca signal. The fluorescence signal was detected using a single element silicon drift diode. Additional micro-XRF maps of the heavy metals (e.g., Ni, Co) were collected at the Advanced Light Source (ALS) on beamline 10.3.2 in Berkeley, USA (30). The micro-XRF maps were obtained at the energy of 10 keV with a beam size of 5x5 µm^2 using a 7 element Ge-solid state detector. The micro-XRF measurements were performed with a Si(111) crystal monochromator and were carried out in air at room temperature.

Micro-XAS spectra at the K-edge of Ni (8.333 keV), Co (7.709 keV) and Cr (5.989 keV) were collected on beamline 10.3.2 (ALS) with the same set up as for the micro-XRF maps. The XAS spectra of Cr(III) and Cr(VI) reference solutions (Cr(NO$_3$)$_3$ and K$_2$CrO$_4$ solutions), were also collected on beamline 10.3.2 (ALS). Several bulk-XAS spectra of Ni and Co reference compounds were collected on the Swiss Norwegian Beam Line (SNBL) at the European Synchrotron Radiation Facility (ESRF) in Grenoble, France and used as supporting information to identify Ni and Co speciation in the cement matrix (35,36).

Micro-XAS spectra at the Al K-edge (1.559 keV) were recorded on the LUCIA beamline (SLS) with a YB66 crystal monochromator and a beam size of 10x10 µm^2. The XAS spectra of several reference compounds, i.e. ettringite (Ca$_6$Al$_2$(SO$_4$)$_3$(OH)$_{12}$·26H$_2$O) (37), calcium-monosulfoaluminate (3CaO·Al$_2$O$_3$·CaSO$_4$·12H$_2$O), calcium-monocarbo-aluminate (3CaO·Al$_2$O$_3$ CaCO$_3$·11H$_2$O), aluminate (C$_3$A, Ca$_3$Al$_2$O$_6$) and tetra-calcium aluminate hydrated (C$_4$AH$_{13}$, 4CaO·Al$_2$O$_3$ 13H$_2$O) (the latter four compounds from 38) were collected during the same measuring campaign and used to identify the Al speciation in the cement matrix.

The micro-XRD measurements were performed on the beamline 10.3.2 (ALS) with a beam size of 5x5 µm^2 at the energy of 17 keV using a Bruker CCD camera. Experimental parameters such as sample-detector distance, detector plane orientation, etc., were refined using an Al_2O_3 powder material as reference.

2.2.3.2 Micro-XAS data reduction

Data reduction was performed using the WinXAS 3.1 software package following standard procedures (e.g., 39). The spectra were normalized by fitting a first-degree polynomial to the pre-edge and a third-degree polynomial to the post-edge regions. The energy was converted to photoelectron wave vector units ($Å^{-1}$) by assigning the origin E_0 to the first inflection point of the absorption edge. Radial Structure Functions (RSF) were obtained by Fourier transforming the k^3-weighted $\chi(k)$ functions between 3.2 and 10.9 $Å^{-1}$ with a Bessel window function with a smoothing parameter of 4. Multishell fits were performed in real space across the range of the first two shells (ΔR=0.8-3.5 Å). Theoretical scattering paths for the fit were calculated using FEFF 8.20 (40) and the structure of β-Ni(OH)$_2$, Co(OH)$_2$ and CoOOH as references. The amplitude reduction factor (S_0^2) was determined to be 0.85 for both Ni (41) and Co (42). Errors on the structural parameters were estimated from the analysis of a series of reference compounds (see Table 2.2 and Table 2.3).

2.3 Results and discussion

2.3.1 The role of cement in waste disposal

Cement-based materials play an important role in multi-barrier concepts developed worldwide for the safe disposal of hazardous and radioactive wastes (e.g., 43,44). In the case of the latter waste forms, cement is used to condition and stabilize the waste materials and to construct the engineered barrier systems (container, backfill and liner materials) of repositories for radioactive waste. For example, ~95 wt% of the near-field material of the planned Swiss disposal cavern for low and intermediate level waste consists of hardened cement paste (HCP) and cementitious backfill materials.

The immobilization potential of HCP originates from the selective binding properties for different chemical elements (e.g., 45), indicating that retention in cement systems is highly specific with respect to the mineral components and processes involved (e.g.,15, 41,46-48). From a chemical standpoint cement is a very heterogeneous material with discrete particles in the nano- to micrometer size range. Neglecting the micro-heterogeneity of such a complex matrix could result in misunderstanding of uptake processes. Therefore, microscopic investigations were conducted to address the complexity of the metal enriched cementitious matrices used.

2.3.2 Microscopic investigation

The combination of BSE imaging with EDS micro-analysis allows spatially resolved semi-quantitative information on the chemical composition of the different mineral phases in heterogeneous materials to be obtained. SEM-based BSE and EDS have proven to be suitable tools for investigating the microstructure of cement, the spatial distribution of key and major elements in cement and the correlations among elements present in the cement matrix (e.g., 3,5). Fig. 2.1 shows the respective images of the Ni enriched hydrated cement material. Fig. 2.1a illustrates that hydrated cement is a complex mixture of non-hydrated clinker minerals and hydrate assemblages. Non-hydrated cement is mainly composed of alite (A; Ca_3SiO_5), belite (B; Ca_2SiO_4) together with smaller percentages of aluminate (C_3A, $Ca_3Al_2O_6$), ferrite (F; $Ca_4Al_2Fe_2O_{10}$) and anhydrate ($CaSO_4$) (49,50). The presence of non-hydrated clinker minerals in a hydrated cement matrix is illustrated in Fig. 2.1b. During hydration, which starts upon addition of water to non-hydrated cement, the clinker minerals dissolve to form the hydrate assemblage. For example, alite and belite decompose to form calcium silicate hydrates (C-S-H) and portlandite ($Ca(OH)_2$) (Fig. 2.1c and d). C-S-H is the most abundant component of the hydrated assemblage and precipitates as inner-C-S-H and outer-C-S-H. Inner-C-S-H gradually fills up the space originally occupied by alite grains. After prolonged hydration time, most of the alite grains are surrounded by rims of inner-C-S-H, which vary in thickness (i-C-S-H; Fig. 2.1a and c). Outer-C-S-H, on the contrary, fills the pore space of the hydrated cement, originally occupied by water (o-C-S-H; Fig. 2.1a and c) (5,50). Outer-C-S-H is generally less dense than inner-C-S-H and finely intermixed with portlandite (P; Fig. 2.1a and d) and several other minor hydrated phases, such as ettringite ($Ca_6Al_2(SO_4)_3(OH)_{12}\cdot 26H_2O$) and calcium-monosulfoaluminate ($3CaO\cdot Al_2O_3\cdot CaSO_4\cdot 12H_2O$). Depending on the composition of the non-hydrated cement, further important Al-containing hydrated phases form during hydration such as calcium-monocarboaluminate ($3CaO\cdot Al_2O_3 CaCO_3\cdot 11H_2O$),

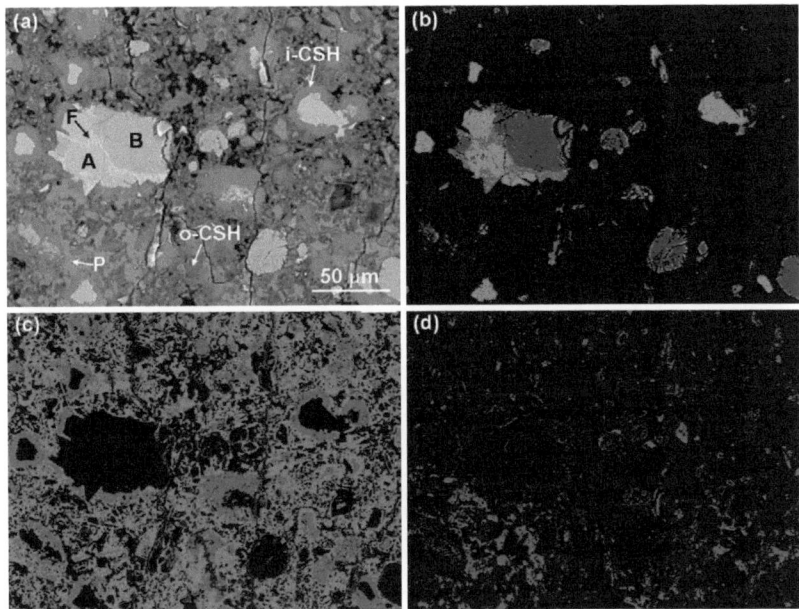

Fig. 2.1. a) BSE-image of the cement matrix hydrated for 30 days. Notations: F = ferrite ($Ca_4Al_2Fe_2O_{10}$), A = alite (Ca_3SiO_5), B = belite (Ca_2SiO_4), P = portlandite ($Ca(OH)_2$), o-C-S-H = outer-Ca-Si-hydrate, i-C-S-H = inner-Ca-Si-hydrate. Figures b-d show the same area as in (a) with mineral phase enhancement. b) shows the phases of ferrite in red, alite in green and belite in blue; c) Outer-C-S-H in green and inner-C-S-H in red; d) Portlandite in red.

tetracalcium aluminate hydrate (C_4AH_{13}, $3CaO \cdot Al_2O_3 \cdot 13H_2O$) and hydrotalcite ($Mg_4Al_2(OH)_{12} \cdot CO_3 \cdot 2H_2O$) (e.g., 49,51). The complexity of the cement matrix is further reproduced by the BSE image and EDS maps shown in Fig. 2.2. Highly concentrated regions of Ca indicate the presence of portlandite and alite. Portlandite only contains Ca, and, therefore, anti-correlates with all other elements. In contrast, alite is a Ca-Si oxide and, therefore, positive correlation between Ca and Si is oberved for this phase. The lower concentrated Ca regions also reveal a positive correlation with Si, which indicate the presence of inner-C-S-H and outer-C-S-H. EDS-point analyses support these findings: alite is marked by higher Ca content (~41-49 wt%; Table 2.1, see analyses a and b shown in Fig. 2.2) than C-S-H (~38 wt%; Table 2.1, see analysis c shown in Fig. 2.2), and a higher Ca/Si ratio (3:1), in accordance with the stochiometric formula (50). An additional phase, which is highly enriched in Al, Fe and some Ca, particularly visible at a single spot, is ferrite and/or hydrated products of ferrite.

Table 2.1. Semi-quantitative EDS-analyses for single regions. All data are given in wt% (spots are indicated in Fig. 2.2).

	Ca	Si	Fe	Al	S	Ni	Mg	Cr	Sr	K	Na	Cl	O	Total
a	48.80	10.45	0.14	0.26	0.15	0.01	0.27	0.00	0.00	0.06	0.16	0.03	39.69	40.22
b	41.13	10.72	1.12	1.05	1.07	0.60	0.25	0.05	0.00	0.08	0.00	0.00	43.95	44.93
c	37.83	11.11	0.68	0.45	0.90	0.00	0.51	0.09	0.00	0.00	0.12	0.05	48.26	49.03
d	25.63	8.43	0.48	1.92	0.79	10.85	0.43	0.00	1.96	0.01	0.08	0.03	49.41	62.77
e	38.21	8.08	1.07	1.43	0.78	5.97	0.38	0.08	2.02	0.06	0.00	0.08	41.84	50.43
f	24.38	8.49	0.32	2.46	0.70	13.16	0.42	0.02	2.02	0.07	0.00	0.02	47.96	63.67
g	25.81	8.27	0.47	2.29	0.96	12.27	0.30	0.07	2.01	0.02	0.00	0.07	47.45	62.19
h	33.89	6.96	1.16	1.54	0.77	4.23	0.51	0.01	1.70	0.00	0.00	0.05	49.16	55.66
i	26.10	8.21	0.48	1.44	0.58	7.51	0.63	0.05	1.88	0.00	0.28	0.00	52.85	63.20

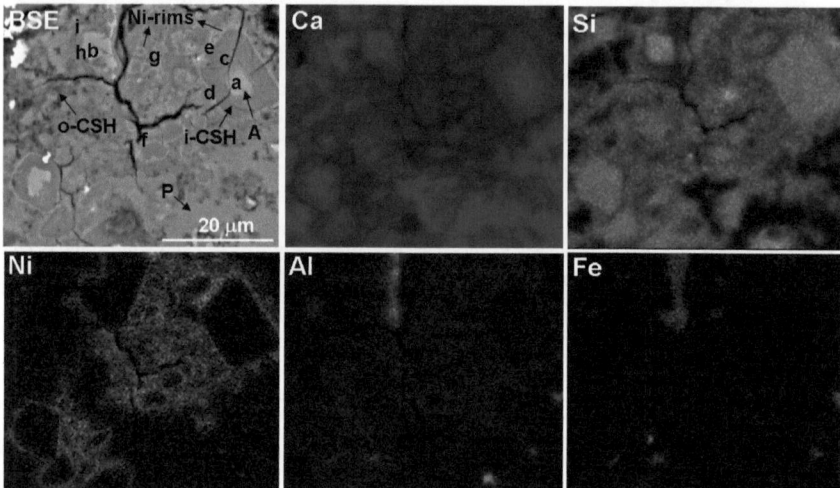

Fig. 2.2. BSE image of a Ni-rich region together with EDS elemental distribution maps of a Ni enriched hydrated cement matrix. Notations: P= portlandite $(Ca(OH)_2)$, i-C-S-H= inner-C-S-H, o-CHS= outer-C-S-H.

In the next step of our investigations we addressed the question of the distribution of Ni in this complex cement matrix, in particular with a view to elucidating its association with specific minerals of the cement matrix. The relevant information also stems from SEM-EDS and is shown Fig. 2.2 for the case of the Ni enriched hydrated cement matrix. The BSE image reveals bright rims around inner-C-S-H phases. The spatially resolved EDS elemental distribution maps show that these bright rims are enriched in Ni. Furthermore, a weak correlation between Ni and Al was found in these rims, which clearly emerges from the EDS point analyses (compare spots d, e, f, g, h and i in Table 2.1). The EDS-point analyses further reveal that the Ni-rich phase has a variable Ni content, ranging from ~5 wt% up to ~10 wt% (Table 2.1). Although the EDS-maps suggest an anti-correlation between Ni and Ca or Si, respectively, the EDS point analyses reveal that the Ni-rich rims contain small amounts of Ca, Si as well as Al. The information gained from the microscopic SEM-BSE/EDS investigations enables us to develop a mechanistic view of the behaviour of Ni in the hydrating cement. The key information is the fact that Ni-rich rims form around alite. A tentative explanation for Ni accumulation around alite can be given by considering some specific features of the mineral. Alite dissolves much faster than the other main clinker mineral, belite, which allows inner-C-S-H type zones to be rapidly formed around alite grains in contrast to outer C-S-H phases (50). These reactive zones are expected to have a high specific surface area. Therefore, a large number of surface sites may be available for Ni binding, which may facilitate Ni accumulation at the grain boundaries of alite. Correlation with Al further suggests that a mixed Ni-Al phase could form. Nevertheless, it clearly appears that SEM-based BSE and EDS do not provide more detailed information on the speciation of such mixed Ni-Al phases present in the hydrated cement matrix. This information is retrieved by using micro-XAS.

2.3.3 Synchrotron-based investigations

2.3.3.1 Micro-XRF investigations

Micro-XRF elemental maps are required in a first step of synchrotron-based speciation studies to gain an overview on the distribution of the investigated chemical elements and to localize Ni enriched spots in the cement matrix, where micro-XAS spectra will be collected. In the ideal case, it is desirable to investigate the same areas of interests or spots, respectively, previously investigated by SEM-based BSE and EDS. This study illustrates that this ideal situation can be achieved.

In order to roughly re-localize an area of interest that was accessible for both techniques (SEM-based BSE/EDS and micro XRF) as well as during the various experimental runs at different beamlines a silver spot (Fig. 2.3, Ag on the BSE image, in blue on the micro-XRF maps, Si, Ca, Al, Ni, S) was implemented as a marker.

Micro-XRF maps of ~700x1000 μm^2 were recorded on the same area previously imaged by BSE (Fig. 2.3). As describe above in more details the BSE image shown in Fig. 2.3 reveals bright grains (clinker minerals), fine darker grains (hydrated minerals), and black zones, which correspond to the pore space in the hydrated cement matrix (52). The region investigated by SEM is strongly hydrated, lies between two clinker minerals and is marked with '1' on the BSE image in Fig. 2.3.

Micro-XRF mapping focused on the elemental distribution of Ni, Si, Ca, Al and S. The latter four elements are among the most abundant elements in the cement matrix. The Si map allows clinker minerals to be identified (yellow regions on the map). The highest concentrated Si region (red) is a quartz grain. The latter finding was confirmed by BSE imaging (details with high magnification are not shown). The Ca distribution, on the contrary, mainly reflects zones with hydrated cement minerals. The regions coloured in red are highly concentrated Ca regions, indicating the presence of mainly inner-C-S-H and clinker minerals. Regions with less Ca in green indicate the formation of outer-C-S-H and portlandite. The knowledge of the S distribution in combination with the Al map allows regions of high Al and high S concentrations to be distinguished from regions with high Al but low S. The latter case indicates the presence of non-hydrated aluminate, whereas areas with high Al and S concentrations indicate the presence of ettringite and/or calcium-monosulfoaluminate, (51). The combination of Ni and Al distribution maps is needed to localize the areas of interest for chemical speciation of Ni using micro-XAS. Note that it was postulated that mixed Ni-Al phases may form in cementitious materials (see previous section '*microscopic investigation*') (41). With the above given information, the region of interest (spot 1) for Ni speciation studies can be easily retrieved by the synchrotron-based X-ray micro-beam studies.

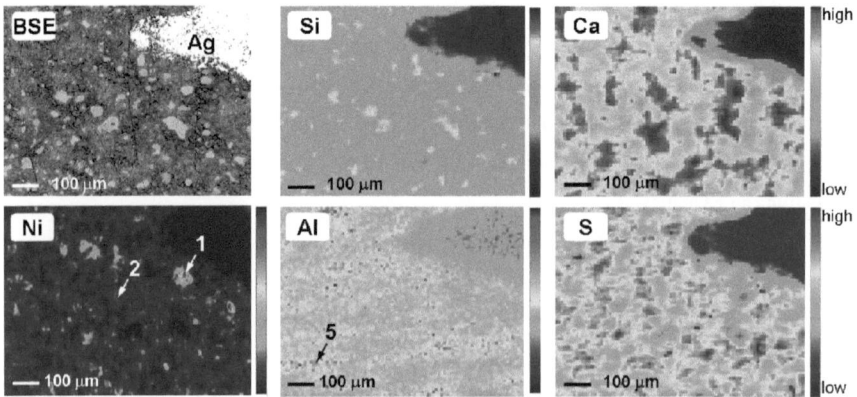

Fig. 2.3. BSE image and micro-XRF elemental distribution maps of Si, Ca, Al and S in a 700x1000 μm^2 overview of a Ni enriched hydrated cement matrix. The region marked with 1

corresponds to the same region illustrated in Fig. 2.2. Micro-XAS selected regions for Ni K-edge (1, 2) and for Al K-edge (5) measurements are marked with numbers on the BSE image and the respective elemental maps. Ag = silver spot used as marker, which appears blue on the micro-XRF maps.

It is worth noting that there are some differences between the fluorescence signal detected by SEM-EDS and by synchrotron-based micro-XRF measurements. The higher penetration depth (the absorption length of X-rays within a cement matrix is ~150 μm at ~8 keV) and the lower spatial resolution (few micron, see *'materials and methods'*) of the synchrotron X-rays compared to those of the electrons (penetration depth ~1 μm and spatial resolution: nanometer range) results in the representation of spot-like structures on the micro-XRF maps (e.g. Ni map, Fig. 2.3) instead of rims as shown on the SEM-EDS maps (e.g. Ni map, Fig. 2.2). Nevertheless, the approach used in this study (combined use of SEM and micro-XRF) is sufficient to re-localize an area of interest in the cement matrix and perform spatially-resolved micro-XAS investigations for chemical speciation, the key to molecular-level information. In the following sections results from speciation studies on Ni-, Co- and Cr-enriched hydrated cement will be discussed to illustrate the potential of micro-XAS to gain information on the chemical speciation of a given element of interest in a highly heterogeneous material.

2.3.3.2 Ni speciation in Ni(II) enriched hydrated cement

The first example illustrates our approach towards a better understanding of the chemical speciation of Ni in Ni(II) enriched hydrated cement using micro-XAS (52). In Fig. 2.4 selected micro-EXAFS measurements of single Ni-rich and Ni-poor spots (marked with 1 and 2 in Fig. 2.3) and relevant Ni reference spectra are presented. Fig. 2.4a shows the normalized, background-subtracted and k^3-weighted micro-EXAFS spectra.

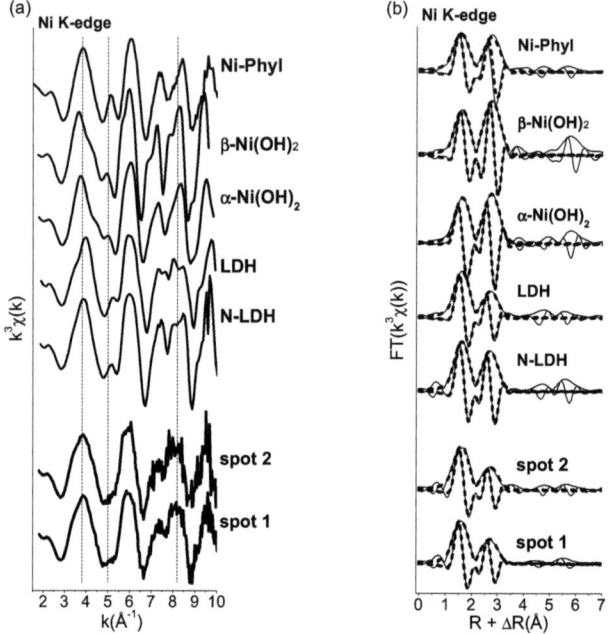

Fig. 2.4. Ni K-edge bulk-EXAFS experimental spectra of reference compounds and micro-EXAFS of a Ni enriched hydrated cement sample (5000 mg/kg metal loading, spot 1 and 2 are illustrated in Fig. 2.3.); a) k^3-weighted, normalized, background-subtracted spectra; b) experimental (solid line) and theoretical (dashed line) Fourier Transforms (modulus and imaginary parts) obtained from the spectra presented in Fig. 2.4a. Notations: N-LDH = neo-formed Ni-Al LDH (53), LDH = synthetic Ni-Al LDH (Ni:Al, 2:1), Ni-Phyl = Ni-phyllosilicate (14).

All micro-EXAFS spectra performed on single individual spots reveal similar features irrespective of the Ni concentration on the spot. This finding suggests the presence of a similar coordination environment of Ni, independently from the Ni concentration. A closer inspection of the spectra shows that the oscillation at ~4 Å^{-1} is broad and is located at a k-range that agrees with the position of the first major oscillation observed in the spectrum of a neo-formed LDH phase (N-LDH) (53) (Fig. 2.4a). The splitting of the oscillation at ~8 Å^{-1} is a characteristic beat pattern. Scheinost and Sparks demonstrated that this beat pattern indicates the presence of Ni-Al LDH, and can be used as a fingerprint (54). In fact, the beat pattern at ~8 Å^{-1} is observed in both Ni-Al LDH spectra (LDH and N-LDH), whereas the other reference compounds (α-Ni(OH)$_2$, β-Ni(OH)$_2$ and Ni-phyllosilicate) show an elongated upward oscillation ending in a sharp tip at ~8.5 Å^{-1}. Thus, the presence of the beat pattern at ~8 Å^{-1} together with the observed spectral features at ~4 Å^{1} suggest that a Ni-Al LDH phase has formed at single spots (e.g ., spot 1 and 2) in the hydrated cement matrix enriched in Ni. The corresponding Fourier transforms (FT) of the k^3-weighted micro-EXAFS spectra further corroborate this finding (Fig. 2.4b). The positions of the first and second FT peaks, as well as the shape of the imaginary part of the Ni enriched hydrated cement samples do agree well with those from Ni-Al LDH compounds. Note that the amplitude of the second peak in the experimental spectra of spot 1 and spot 2 is clearly reduced compared to the Ni-hydroxide reference spectra. Such an amplitude reduction results from the destructive inference of Ni and Al backscattering contribution (41,46,54, see below) which is also observed in the Ni-Al LDH reference spectra.

The structural parameters derived from multi-shell analysis (ΔR=0.8-3.5 Å) are summarized in Table 2.2. The first coordination shell was fitted with Ni-O backscattering pairs. The second coordination shell was fitted solely using Ni-Ni pairs, as the discrimination of Ni-Ni and Ni-Al backscattering pairs in Ni-Al LDH is problematic (41,46). To be able to compare the coordination numbers of the Ni-Ni backscattering pairs (CN$_{Ni-Ni}$) determined in the cement samples and the references, the Debye-Waller factor was fixed to 0.005 Å^2 as determined for the β-Ni(OH)$_2$. Data analysis reveals similar CN and interatomic distances (R) for both spots (Table 2.2). The first FT peak corresponds to an octahedral coordination of Ni with ~6 oxygen atoms at 2.03-2.06 Å. The second FT peak reveals strongly reduced CN$_{Ni-Ni}$ (~3) compared to α-Ni(OH)$_2$ (~5) (55) and β-Ni(OH)$_2$ (~6) (56). The CN$_{Ni-Ni}$ derived form the cement spectra are comparable to those determined from the Ni-Al LDH spectrum. The CN$_{Ni-Ni}$ is reduced as Ni is partly substituted by Al in Ni-Al LDH. This results in a significant destructive interference between Ni and Al EXAFS contributions and, consequently, in an amplitude cancellation of the Ni and Al shells (46). Although the CN$_{Ni-Ni}$ of the single Ni spots and Ni-Al LDH agree very well, R$_{Ni-Ni}$ of 3.09-3.10 Å indicate that not a pure Ni-Al LDH phase (R$_{Ni-Ni}$=3.06-3.07 Å) has formed in the investigated areas. This finding was attributed to the presence of a small amount of β-Ni(OH)$_2$ (R$_{Ni-Ni}$=3.13 Å) which causes longer Ni-Ni distances. The results gained from the micro-XAS investigations are complementary and in good agreement with the SEM-EDS investigations. The findings revealed that Ni is mainly bound to Al predominantly as Ni-Al LDH phase and only minor amounts of Ni-hydroxides formed (52).

Table 2.2. Structural information obtained from selected micro-EXAFS Ni K-edge data analysis together with reference compounds (spots are indicated in Fig. 2.2).

Samples	Ni-O			Ni-Ni			Ni-Si			ΔE_0	% Res
	CN	R (Å)	σ^2 (Å)	CN	R (Å)	σ^2 (Å)	CN	R (Å)	σ^2 (Å)		
References											
Ni-Phyllosilicate [a]	5.1	2.04	0.006	3.5	3.07	0.008[c]	3.7	3.26	0.008[c]	0.3	3.0
β-Ni(OH)$_2$	5.6	2.06	0.005	5.6	3.13	0.005				-0.6	3.0
α-Ni(OH)$_2$	5.2	2.03	0.005	4.9	3.09	0.005[d]				3.0	4.4
Ni-Al LDH (LDH)	6.0	2.05	0.006	2.5	3.06	0.005[d]				1.1	4.5
Neo-formed Ni-Al LDH [b] (N-LDH)	5.7	2.04	0.004	3.9	3.07	0.005[d]				0.3	3.8
Sample											
Spot 1	5.7	2.04	0.006	2.5	3.09	0.005[d]				-1.9	6.1
Spot 2	6.6	2.06	0.007	2.3	3.10	0.005[d]				-0.4	9.9

[a] Dähn et al. 2002, [b] Scheidegger et al., 1997, [c] correlated parameters and [d] fix parameters during fitting procedures.
R, CN, σ^2 ΔE_0 stand for interatomic distances, coordination numbers, Debye-Waller factors and inner potential corrections.
Estimated error: $R_{(Ni-O)}$ ±0.02 Å, $CN_{(Ni-O)}$ ±20%, $R_{(Ni-Ni)}$ ±0.02 Å, $CN_{(Ni-Ni)}$ ±20%
% Res: deviation between experimental data and fit given by the relative residual in percent.
N= number of data points, Yexp and Ytheo: experimental and theoretical data points, respectively.

$$\% \operatorname{Res} = \frac{\sum_{i=1}^{N} |y_{exp}(i) - y_{theo}(i)|}{\sum_{i=1}^{N} y_{exp}(i)} * 100$$

2.3.3.3 Co speciation in Co(II) enriched hydrated cement

In a subsequent study, the immobilization of Co(II) in hydrated cement was investigated to compare its behaviour with Ni(II). Ni(II) and Co(II) were thought to have a very similar chemical behaviour in cementitious systems, since both are divalent cations. The micro-XRF map shown in Fig. 2.5 reveals the heterogeneous distributions of Co as previously observed in the case of Ni. The maps of two different regions in the hydrated cement matrix show that Co can be enriched in spot-like (Fig. 2.5a) or ring-like (Fig. 2.5b) structures. Micro-EXAFS spectra were collected at two spots (Fig. 2.5c), i.e. spot 3 on the spot-like and spot 4 on the ring-like structure. The normalized, background-subtracted and k^3-weighted micro-EXAFS spectra shown in Fig. 2.5c of the Co(II) enriched hydrated cement matrix reveal clear differences for the two spots. The micro-EXAFS spectrum collected at spot 3 and the EXAFS spectra of Co(II) reference compounds are well in phase. Further similarities are the position of the first oscillation at ~4 Å^{-1}, the position of the oscillations at ~6 Å^{-1} and the k-range between 7-8.5 Å^{-1}. In contrast, the spectrum collected at spot 4 reveals more similarities with the EXAFS spectra of Co(III) reference compounds. Again, the spectra are more or less in phase. Furthermore, the oscillation at ~4 Å^{-1}, at ~7 Å^{-1} and the k-range between ~8-9.5 Å^{-1}

illustrate strong similarities to CoOOH and/or Co-phyllomanganates (Co-asbolane and/or Co-buserite, 18,57). The distinct spectral feature and the clear difference in phase observed

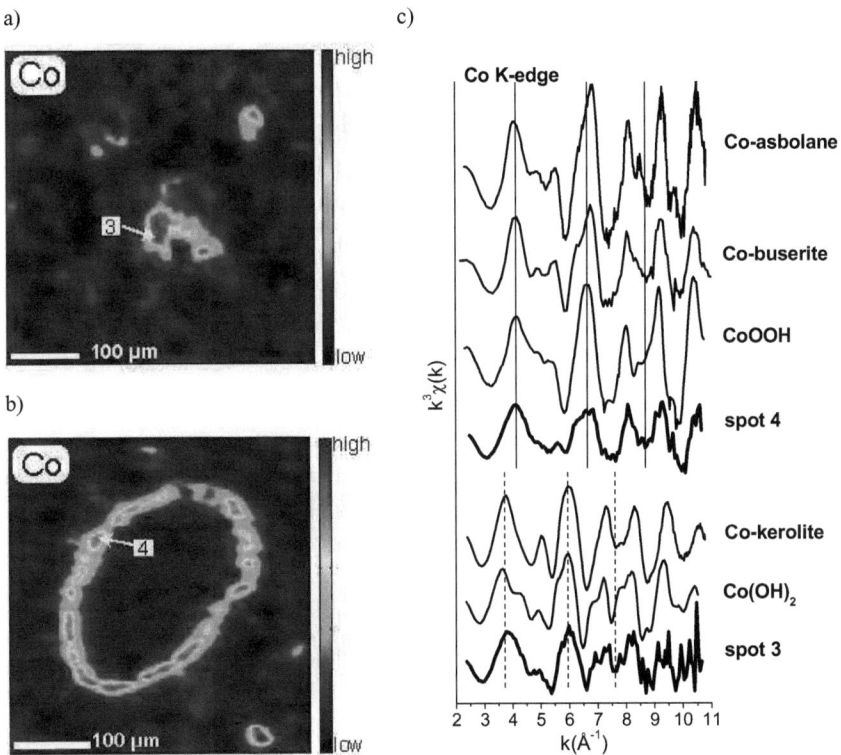

Fig. 2.5. Micro-XRF elemental distribution maps of Co in a Co(II) enriched hydrated cement sample (5000 mg/kg metal loading; a) Spot 3; b) Spot 4; c) Co K-edge k^3-weighted, normalized, background-subtracted micro-EXAFS experimental spectra of the single spots 3 and 4 together with bulk-EXAFS of Co-reference compounds: Co-kerolite (58); Co-asbolane (18); Co-buserite (57), CoOOH and Co(OH)$_2$ (36).

in the k^3-weighted micro-EXAFS spectra among the two spots (Co(II) and Co(III) species) are further reflected in the entirely different distances determined for the first (Co-O) and the second (Co-Co) coordination shells. For spot 3, data analysis indicates Co-O and Co-Co distances of 2.06 Å and 3.16 Å, respectively (Table 2.3). These distances are typical for Co(II) species, in particular Co-hydroxide (57). On the other hand, the data analysis for spot 4 results in first and second shell Co(III) distances (R_{Co-O}=1.90 Å and R_{Co-Co}=2.80 Å) that are similar to Co-O and Co-Co distances in CoOOH and Co-phyllomanganate (Table 2.3) (18,57). The results from the micro-EXAFS show that Co immobilized in the cement matrix during cement hydration is present in two different oxidation states (Co(II), Co(III)). Detailed investigations were carried out to further elucidate the Co oxidation process (36). The results showed that oxygen strongly influences the speciation of Co in hydrating cement. Oxygen concentrations present in the hydrating cementitious system, e.g. either trapped in the cement powder or dissolved in the water used, strongly influences the Co speciation. The results suggested that oxidation of Co(II) to Co(III) in cementitious systems is fast and can only be suppressed by rigorous measures taken to exclude oxygen. The findings from the investigation of the Co speciation illustrate the importance of performing micro-scale chemical speciation, as the Co speciation was found to vary significantly at different regions of the cement matrix. Further, the finding has implications for the Co release from cement-stabilized wastes as the solubility of Co(III)-phases is lower than that of Co(II)-phases.

Table 2.3. *Structural information obtained from selected micro-EXAFS Co K-edge data analysis together with reference compounds (spots are indicated in Fig. 2.5).*

	1st shell						2nd shell							
	Co(II)-O			Co(III)-O			Co(II)-Co(II)			Co(III)-Co(III)				
Samples	CN	R (Å)	σ^2 (Å^2)	CN	R (Å)	σ^2 (Å^2)	CN	R (Å)	σ^2 (Å^2)	CN	R (Å)	σ^2 (Å^2)	ΔE_0 (eV)	%Res
References														
Co(OH)$_2$	4.1	2.09	0.005	-	-	-	5.0	3.16	0.007	-	-	-	-7.3	5.5
Co-Al LDH	6.5	2.09	0.006	-	-	-	2.3	3.09	0.006	-	-	-	0.3	3.0
Co-Kerolite^a	4.4	2.09	0.005	-	-	-	7.2	3.13	0.008	-	-	-	-4.6	6.9
CoOOH	-	-	-	4.0	1.90	0.001	-	-	-	4.6	2.85	0.004	0.7	
Co-Buserite^a	-	-	-	5.9	1.89	0.009	-	-	-	4.6	2.80	0.007	-1.8	27.6
Co-Asbolane^a	-	-	-	5.2	1.89	0.004	-	-	-	7.2	2.80	0.007	-1.2	3.9
Samples														
Spot 3	4.3	2.06	0.009	-	-	-	4.1	3.16	0.009^c	-	-	-	6.2	17.4
Spot 4	-	-	-	5.3	1.90	0.01	-	-	-	1.8	2.80	0.005^b	-2.0	15.3

^{a} Manceau et al., 1997, ^{b} fix parameters during fitting procedures and ^{c} correlated parameters
CN, R, σ^2, ΔE_0 stand for interatomic distance, coordination number, Debye-Waller factor and inner potential correction.
Estimated errors: $R_{(Co-O)}$ ±0.02 Å, $CN_{(Co-O)}$ ±20%, $R_{(Co-Co)}$ ±0.02 Å, $CN_{(Co-Co)}$ ±20%
%Res: deviation between experimental data and fit given by the relative residual in percent.
N= number of data points, Y_{exp} and Y_{theo}: experimental and theoretical data points, respectively.

$$\% \operatorname{Re} s = \frac{\sum_{i=1}^{N}|y_{exp}(i) - y_{theo}(i)|}{\sum_{i=1}^{N} y_{exp}(i)} * 100$$

2.3.3.4 Oxidation state of Cr in Cr(VI) enriched hydrated cement

Determination of the oxidation state of Cr in cementitious materials is particularly important. Cr, especially Cr(VI), is known as a skin sensitizer, which can lead to 'dermatitis'. 'Dermatitis' is the most frequent health disorder among workers who are regularly in contact with cement (59-61). Wet cement has a relatively high pH (pH>12.5), which can alter the outermost layer of the skin, facilitating the penetration of water-soluble substances. The solubility of Cr strongly depends on the valence state. Cr(VI) is more soluble, thus more mobile than Cr(III), exhibiting different behaviour in the cementitious matrix. Therefore, it is important to accurately determine the Cr(VI)/(III) speciation and amounts present in cementitious matrices.

The identification of Cr(VI) and Cr(III) occurs based on the pre-edge feature, which is pronounced in the case of Cr(VI), but very small for Cr(III) (Fig. 2.6a) (62). The pre-edge feature is an electronic transition from a core level to the lowest unoccupied or partially occupied energy level. In the case of the K-edge of Cr(VI) the pre-edge feature is due to a 1s -> 3d transition (63).

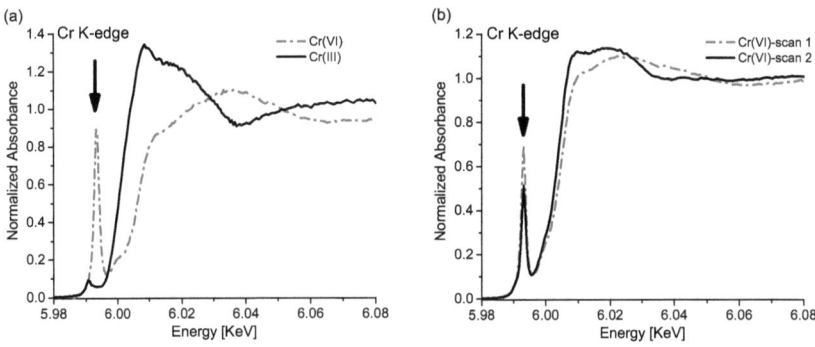

Fig. 2.6. Cr K-edge micro-XANES spectra; a) reference spectra of K_2CrO_4 and $Cr(NO_3)_3$ solutions used for identifying Cr(VI) and Cr(III) species, respectively; b) experimental spectra obtained from a thin section of Cr(VI) enriched cement (5000 mg/kg metal loading). The spectrum indicated with a dashed line is the first spectrum collected under the X-ray beam. The spectrum illustrated with a solid line was collected after ~20 minutes exposure time on the same spot. Arrows indicate pre-edge feature.

Although, the identification of Cr(VI) and Cr(III) is straightforward, micro-XAS studies with Cr on cementitious materials were found to be problematic. During the experiments conducted at 10.3.2 (ALS) it was observed that the oxidation state of Cr(VI) changed under the micro-beam (beam size of 5x5 μm^2). CrO_4^{2-} (Cr(VI)) was found to reduce to Cr(III) within ~20 minutes after irradiation started. Fig. 2.6b shows the first scan of a Cr(VI) enriched hydrated cement matrix on a region of interest rich in Cr(VI). The second scan was collected after ~20 minutes exposure time, showing the reduction of Cr(VI) to Cr(III). These investigations clearly demonstrated that, although XAS is considered to be a non-destructive technique, in some cases, special care has to be taken to exclude artefacts caused by beam damage. Neglecting such phenomena could lead to a misinterpretation of the XAS data and consequently give rise to conflicting information on immobilization processes involved. Thus, the lesson learned from this study is the following: when investigating the speciation of redox sensitive elements with highly brilliant beams (high photon flux within small beam area), it is required to compare the micro-XAS results of the first and last scan of a series of scans taken from the area of interest. The intense radiation provided by state of the art synchrotron facilities may trigger electrochemical processes and change the oxidation state of the elements under investigation. Beam-induced redox processes of redox sensitive elements, such as Mn and Cr, have been reported in several studies (62,64). Yet, no systematic study dealing with the factors causing beam damage and underlying mechanisms of such beam damages have been performed to date.

2.3.3.5 Al speciation in hydrated cement

Micro-XAS investigations further allow determining the chemical speciation of elements inherent to the cement matrix. Knowledge of the speciation of cement-derived major elements (Ca, Si, S and Al) may provide complementary information to that gained for the impurity elements, such as Ni, Co and Cr, which were added to hydrating cement. The former information is particularly important if specific cement minerals are involved in the binding mechanism of a specific impurity element. An important aspect of the micro-XAS studies with the lighter cement-derived elements such as Ca, Si, S and Al is that these investigations have to be carried out at lower energies of the X-ray beam (< ~5 keV) in a vacuum or a He environment. Note that the elements Ni, Co, and Cr have K-edges at energies significantly higher than ~5 keV. At the latter energies, attenuation of the X-rays by air absorption is low, and thus, no special conditions are needed for the sample environment.

Investigations of Al speciation in cement were carried out at the LUCIA beamline (SLS, Switzerland). Due to the overlap with Si (Si K-edge: 1839 eV or only 280 eV above the Al K-edge), Al K-edge (1559 eV) XAS experiments with cement provide only data of a limited spectral range (XANES data).

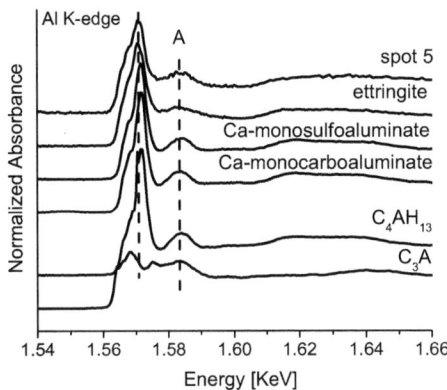

Fig. 2.7. Al K-edge micro-XANES experimental spectra of hydrated cement and powder experimental spectra for reference compounds: ettringite ($Ca_6Al_2(SO_4)_3(OH)_{12}\cdot 26H_2O$), calcium-monosulfoaluminate ($3CaO\cdot Al_2O_3\cdot CaSO_4\cdot 12H_2O$), calcium-monocarboaluminate ($3CaO\cdot Al_2O_3\ CaCO_3\cdot 11H_2O$), aluminate ($C_3A$, $Ca_3Al_2O_6$) and tetracalcium aluminate hydrate (C_4AH_{13}, $4CaO\cdot Al_2O_3\ 13H_2O$). The dashed lines indicate the spectral features explained in detail in the text.

The elemental distribution map of Al (Fig. 2.3) illustrates the heterogeneous distribution of Al in the cement matrix with Al-rich and -poor regions. Some of the Al-rich regions show a positive correlation with S, indicating the presence of S-containing cement minerals. This is, for example, the case at spot 5, where the micro-XANES data was collected. Fig. 2.7a shows the micro-XANES of the selected Al-rich spot (marked as spot 5 in Fig. 2.3) and the corresponding Al-containing reference materials. The micro-XANES experimental spectrum of spot 5 on the Al K-edge reveals strong similarities with that of ettringite ($Ca_6Al_2(SO_4)_3(OH)_{12}\cdot 26H_2O$), in particular the position of the absorption edge and the spectral feature 'A'. Further micro-XANES measurements on the Al K-edge were carried out on Ni-rich spots. However, the Al concentration at these spots was to low to obtain analyzable data and, thus, to deduce the Al speciation.

The feasibility study on Al speciation in cement discussed above demonstrates that micro-XAS is also a versatile analytical tool for chemical speciation of light elements (e.g., Al, S, P, Mg, Na). No doubt, with the construction of micro-XAS facilities in a lower energy regime, the technique will become increasingly important to address questions related to spatially-resolved chemical speciation of key elements in environmental and biological systems and to obtain a molecular-level understanding of the physiochemical processes involved.

2.3.4 Synchrotron-based micro-XRD: An upcoming method for mineral phase identification in heterogeneous materials on the micro-level

As mentioned earlier, XAS is a short-range analytical probe, yielding information about the local structure within a distance of ~5 Å around the X-ray absorbing atom of interest. This allows amorphous or poorly crystalline cement phases to be investigated. In many cases, however, it is essential to identify the micro-crystalline minerals in heterogeneous materials using micro-XRD. This is the reason why the interest in combining micro-XAS investigations with micro-XRD measurements on modern synchrotron-based micro-beam facilities is steadily growing. The combined use of micro-XRF, micro-XAS and micro-XRD allows complementary information on elemental distribution, speciation and the crystal structure of mineral phases to be gained on the same sample and on the same micro-scale area of interest. Several micro-focusing beamlines are equipped with imaging detectors of one kind or another for micro-XRD measurements, e.g., 10.3.2 at ALS and micro-XAS at SLS. The following preliminary results from micro-XRD experiments are intended to illustrate the potential of the method. Fig. 2.8 shows micro-XRD images collected on the thin section at the same Ni-rich and Ni-poor spots (spot 1 and spot 2, Fig. 2.3 and Fig. 2.4) of the Ni enriched hydrated cement that were previously investigated using micro-EXAFS. The micro-XRD images reveal that the cement matrix at the given spot is highly crystalline as indicated by individual spots appearing on diffraction image. It should be noted that, at the present time, further analysis of the data obtained by micro-XRD performed on thin sections is complicated due to the random orientation of the irradiated crystals. The unknown factors are the orientation of the crystals relative to the coordinate frame of the diffractometer and possibly also the crystal lattice itself, i.e. when solving a new structure. At present, the standard method to overcome the problem of crystal orientation is to perform an angular averaging of the diffraction image. This procedure effectively reduces the diffraction image made up of individual spots to a spherically averaged powder pattern (e.g., 33 and references therein). Obviously, this represents a serious loss of information contained in the original images. However, this approach allows a comparison to be made with data in standard crystallographic databases for powder patterns. It is anticipated that, with further development in the field of micro-XRD, the analysis of complex micro-XRD patterns consisting of several individual grains will be possible in the future.

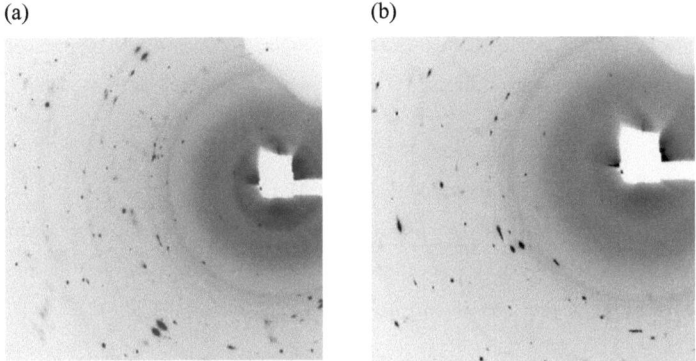

Fig. 2.8. Micro-XRD images collected on the same Ni micro-EXAFS areas shown in Fig. 2.3 and Fig. 2.4; a) Ni-rich spot 1; b) lower Ni concentrated spot 2.

2.4 Conclusions

The multi-technique approach presented in this paper illustrates that the combination of synchrotron-based micro-XRF, micro-XAS and micro-XRD with electron microscopic studies (BSE, EDS) allows spatially-resolved in situ micro-scale investigations of heterogeneous matrices to be carried out. Each of the above presented technique (micro-XRF, micro-XAS, micro-XRD, BSE, EDS) has unique capabilities. BSE and EDS allows spatially-resolved information on the microstructure and elemental distribution of heterogeneous matrices to be discerned. Additionally, micro-XRF combined with micro-XAS allows the chemical speciation of elements of interest on the same sample and on the same area of interest to be determined. Thus, the combination of these techniques shed light from various perspectives on the properties of highly heterogeneous systems and the reaction process therein on the micro-scale.

2.5 References

(1) Catalano, J. G.; McKinley, J. P.; Zachara, J. M.; Heald, S. M.; Smith, S. C.; Brown Jr., G. E. Changes in uranium speciation through a depth sequence of contaminated Hanford sediments. *Environmental Science and Technology* **2006**, *40* (8), 2517-2524.

(2) Famy, C.; Scrivener, K. L.; Crumbie, A. K. What causes differences of C-S-H gel grey levels in backscattered electron images? *Cement and Concrete Research* **2002**, *32* (9), 1465-1471.

(3) Scrivener, K. L. Backscatter electron imaging of cementitious microstructures: understanding and quantification. *Cement and Concrete Composites* **2004**, *26*, 935-945.

(4) Kirpichtchikova, T.; Manceau, A.; Spadini, L.; Panfili, F.; Marcus, M.; Jacquet, T. Speciation and solubility of heavy metals in contaminated soil using X-ray microfluorescence, EXAFS spectroscopy, chemical extraction, and thermodynamic modelling. *Geochimica et Cosmochimica Acta* **2006**, *70*, 2163-2190.

(5) Famy, C.; Scrivener, K. L.; Atkinson, A.; Brough, A. R. Effects of an early or a late heat treatment on the microstructure and composition of inner C-S-H products of Portland cement mortars. *Cement and Concrete Research* **2002**, *32* (2), 269-278.

(6) Egerton, R. F.; Malac, M. EELS in the TEM. *Journal of Electron Spectroscopy and Related Phenomena* **2005**, *143*, 43-50.

(7) Mondi, C.; Leifer, K.; Mavrocordatos, D.; Perret, D. Analytical electron microscopy as a tool for accessing colloid formation process in natural waters. *Journal of Microscopy* **2002**, *207*, 180-190.

(8) Egerton, R. F.; Li, P.; Malac, M. Radiation damage in the TEM and SEM. *Micron* **2004**, *35*, 399-409.

(9) Koningsberger, D. C.; Prins, R. *X-ray absorption*; John Wiley & Sons: New York, **1987**.

(10) Brown Jr., G. E. Spectroscopic studies of chemisorption reaction mechanisms at oxide-water interfaces. In: *Mineral-Water Interface Geochemistry*; Hochella, M. F., White, A. F., Eds.: Washington, DC., **1990**.

(11) Charlet, L.; Manceau, A. Structure, formation, and reactivity of hydrous oxide particles: Insights from X-ray absorption spectroscopy. In: *Environmental particles*; Buffle, P. J., van Leeuwen, H. P., Eds.; Lewis Publishers: Boca Raton, **1993**.

(12) Fenter, P. A.; Rivers, M. L.; Sturchio, N. C.; Sutton, S. R. *Applications of synchrotron radiation in low-temperature geochemistry and environmental science*; The Mineralogical Society of America: Washington DC, **2002**.

(13) Scheidegger, A. M.; Strawn, D. G.; Lamble, G. M.; Sparks, D. L. The kinetics of mixed Ni-Al hydroxide formation on clay and aluminum oxide minerals: A time-resolved XAFS study. *Geochimica et Cosmochimica Acta* **1998**, *62* (13), 2233-2245.

(14) Dähn, R.; Scheidegger, A. M.; Manceau, A.; Schlegel, M. L.; Baeyens, B.; Bradbury, M. H.; Morales, M. Neoformation of Ni phyllosilicate upon Ni uptake on montmorillonite: A kinetics study by powder and polarized extended X-ray absorption fine structure spectroscopy. *Geochimica et Cosmochimica Acta* **2002**, *66* (13), 2335-2347.

(15) Bonhoure, I.; Wieland, E.; Scheidegger, A. M.; Ochs, M.; Kunz, D. EXAFS study of Sn (IV) immobilization by hardened cement paste and calcium silicate hydrates. *Environmental Science and Technology* **2003**, *37*, 2184-2191.

(16) Schlegel, M. L.; Manceau, A.; Chateigner, D.; Charlet, L. Sorption of metal ions on clay minerals. I. Polarized EXAFS evidence for the absorption of Co on the edges of hectorite particles. *Journal of Colloid and Interface Science* **1999**, *215*, 140-158.

(17) Ziegler, F.; Scheidegger, A. M.; Johnson, C. A.; Dähn, R.; Wieland, E. Sorption mechanisms of zinc to calcium silicate hydrate: X-ray absorption fine structure (XAFS) investigation. *Environmental Science and Technology* **2001**, *35*, 1550-1555.

(18) Manceau, A.; Llorca, S.; Galas, G. Crystal chemistry of cobalt and nickel in lithiophorite and asbolane from New Caledonia. *Geochimica et Cosmochimica Acta* **1987**, *51*, 105-113.

(19) Catalano, J. G.; Brown Jr., G. E. Uranyl adsorption onto montmorillonite: Evaluation of binding sites and carbonate complexation. *Geochimica et Cosmochimica Acta* **2005**, *69* (12), 2995-3005.

(20) Reich, T.; Moll, H.; Denecke, M. A.; Geipel, G.; Bernhard, G.; Nitsche, H.; Allen, P. G.; Bucher, J. J.; Kaltsoyannis, N.; Edelstein, N. M.; Shuh, D. K. Characterization of hydrous uranyl silicate by EXAFS. *Radiochimica Acta* **1996**, *74*, 219-223.

(21) Rivers, M. L.; Sutton, S. R.; Smith, J. V. A synchrotron X-ray-fluorescence microprobe. *Chemical Geology* **1988**, *70*, 179.

(22) Vanlangevelde, F.; Tros, G. H. J.; Bowen, D. K.; Vis, R. D. The synchrotron radiation microprobe at the SRS, Daresbury (UK) and is applications. *Nuclear Instruments & Methods in Physics Research Section B-Beam Interactions with Materials and Atoms* **1990**, *49*, 544–550.

(23) Devoti, R.; Zontone, F.; Tuniz, C.; Zanini, F. A synchrotron radiation microprobe for X-ray-fluorescence and microtomography at Elettra – focusing with bent crystals. *Nuclear Instruments & Methods in Physics Research Section B-Beam Interactions with Materials and Atoms* **1991**, *54*, 424-428.

(24) Janssens, K.; Vincze, L.; Adams, F.; Jones, K. W. Synchrotron radiation-induced X-ray-microanalysis. *Analytica Chimica Acta* **1993**, *283*, 98–110.

(25) Hayakawa, S.; Goto, S.; Shoji, T.; Yamada, E.; Gohshi, Y. X-ray microprobe system for XRF analysis and spectroscopy at SPring-8 BL39XU. *Journal of Synchrotron Radiation* **1998**, *5*, 1114–1116.

(26) Newville, M.; Sutton, S.; Rivers, M.; Eng, P. Micro-beam X-ray absorption and fluorescence spectroscopies at GSECARS: APS beamline 13ID. *Journal of Synchrotron Radiation* **1999**, *6*, 353–355.

(27) Bohic, S.; Simionovici, A.; Snigirev, A.; Ortega, R.; Deves, G.; Heymann, D.; Schroer, C. G. Synchrotron hard X-ray microprobe: Fluorescence imaging of single cells. *Applied Physical Letters* **2001**, *78*, 3544–3546.

(28) Somogyi, A.; Drakopoulos, M.; Vincze, L.; Vekemans, B.; Camerani, C.; Janssens, K.; Snigirev, A.; Adams, F. ID18F: A new micro-Xray fluorescence end-station at the European Synchrotron Radiation Facility (ESRF): preliminary results. *X-Ray Spectrometry* **2001**, *30*, 242–252.

(29) Autho In *AIP Conference Proceedings*: USA, **2004**.

(30) Marcus, M.; MacDowell, A. A.; Celestre, R.; Manceau, A.; Miller, T.; Padmore, H. A.; Sublett, R. E. Beamline 10.3.2 at ALS: A hard- X-ray microprobe for environmental and material sciences. *Journal of Synchrotron Radiation* **2004**, *11*, 239-247.

(31) Scheidegger, A. M.; Grolimund, D.; Harfouche, M.; Willimann, M.; Meyer, B.; Dähn, R.; Gavillet, D.; Nicolet, M.; Heimgartner, P. The micro-XAS beamline at the Swiss Light source (SLS): A new analytical facility suited for X-ray micro-beam investigations with radioactive samples. *NEA Publication* **2006**, in press.

(32) Bertsch, P. M.; Hunter, D. B. Applications of synchrotron-based X-ray microprobes. *Chemical Reviews* **2001**, *101* (6), 1809-1842.

(33) Manceau, A.; Marcus, M.; Tamura, N. Quantitative speciation of heavy metals in soils and sediments by Synchrotron X-ray techniques. In: *Application of synchrotron radiation in low-temperature geochemistry and environmental science*; Fenter, P. A., Rivers, M. L., Sturchio, N. C., Sutton, S. R., Eds.; Mineralogical Society of America: Washington, DC, **2002**.

(34) Sutton, S. R.; Bertsch, P. M.; Newville, M.; Rivers, M.; Lanzirotti, A. Microfluorescence and microtomography analyses of heterogeneous earth and environmental materials. In: *Applications of Synchrotron Radiation in Low-Temperature Geochemistry and Environmental Sciences*; American Mineralogical Society, **2002**.

(35) Vespa, M.; Dähn, R.; Grolimund, D.; Wieland, E.; Scheidegger, A. M. Spectroscopic investigation of Ni speciation in hardened cement paste. *Environmental Science and Technology* **2006**, *40*, 2275-2282.

(36) Vespa, M.; Dähn, R.; Grolimund, D.; Wieland, E.; Scheidegger, A. M. Co speciation in hardened cement paste: A macro- and micro-spectroscopic investigation. *Environmental Science and Technology* **2006**, submitted.

(37) Ochs, M.; Lothenbach, B.; Giffaut, E. Uptake of oxo-anions by cements through solid-solution formation: Experimental evidence and modelling. *Radiochimica Acta* **2002**, *90*, 639-646.

(38) Matschei, T.; Lothenbach, B.; Glasser, F. P. The AFm-phase in Portland Cement. *Cement and Concrete Research* **2006**, submitted.

(39) Ressler, T. WinXAS: A program for X-ray absorption spectroscopy data analysis under MS-Windows. *Journal of Synchrotron Radiation* **1998**, *5* (2), 118-122.

(40) Rehr, J. J.; Albers, R. C. Theoretical approaches to X-ray absorption fine structure. *Reviews of Modern Physics* **2000**, *72* (3), 621-653.

(41) Scheidegger, A. M.; Wieland, E.; Dähn, R.; Spieler, P. Spectroscopic evidence for the formation of layered Ni-Al double hydroxides in cement. *Environmental Science and Technology* **2000**, *34*, 4545-4548.

(42) O' Day, P. A.; Brown, J. G. E.; Parks, G. A. X-ray absorption spectroscopy of cobalt (II) multinuclear surface complexes and surface precipitates on kaolinite. *Journal of Colloid and Interface Science* **1994**, *165* (269-289).

(43) Schmidt, M.; Beckefeld, P.; Götz, R.; Kamsties, S.; Kretz, C.; Molitor, N.; Neck, U.; Vogel, P. *Reststoff-und Abfallverfestigung. Immobilisierung von Schadstoffen-Recycling-Verbesserung der Deponiefähigkeit*; Expert Verlag: Renningen-Malmheim, **1995**.

(44) Chapman, N.; McCombie, C. *Principles and standards for the disposal of long-lived radioactive wastes*; First ed.; Elsevier Science, Ltd.: Oxford, **2003**.

(45) Glasser, F. P. Chemistry of cement-solidified waste forms. In: *Chemistry and microstructure of solidified waste forms*; Spence, R. D., Ed.; Lewis Publishers: Boca Raton, **1993**.

(46) Scheidegger, A. M.; Wieland, E.; Scheinost, A. C.; Dähn, R.; Tits, J.; Spieler, P. Ni phases formed in cement and cement systems under highly alkaline conditions: An XAFS study. *Journal of Synchrotron Radiation* **2001**, *8*, 916-918.

(47) Bonhoure, I.; Scheidegger, A. M.; Wieland, E.; Dähn, R. Iodine species uptake by cement and CSH studied by K-edge X-ray absorption spectroscopy. *Radiochimica Acta* **2002**, *90*, 647-651.

(48) Rose, J.; Bénard, A.; J., S.; Borschneck, D.; Hazemann, J.-L.; Cheylan, P.; Vichot, A.; Bottero, J.-Y. First insights of Cr speciation in leached portland cement using X-ray spectromicroscopy. *Environmental Science and Technology* **2003**, *37*, 4864-4870.

(49) Lothenbach, B.; Winnefeld, F. Thermodynamic modelling of the hydration of Portland cement. *Cement and Concrete Research* **2006**, *36*, 209-226.

(50) Taylor, H. F. W. *Cement Chemistry*; Second ed.; Thomas Telford: London, **1997**.

(51) Lothenbach, B.; Wieland, E. A thermodynamic approach to the hydration of sulphate-resisting portland cement. *Waste Management* **2006**, *26*, 706-719.

(52) Vespa, M.; Dähn, R.; Gallucci, E.; Grolimund, D.; Wieland, E.; Scheidegger, A. M. Micro-scale investigation of Ni uptake by cement using a combination of scanning electron microscopy and synchrotron-based techniques. *Environmental Science and Technology* **2006**, in press.

(53) Scheidegger, A. M.; Lamble, G. M.; Sparks, D. L. The kinetics of nickel sorption on phyrophyllite as monitored by X-ray absorption fine structure (XAFS) spectroscopy. *Journal de Physique IV France* **1997**, *7* (C2), 773-775.

(54) Scheinost, A. C.; Sparks, D. L. Formation of layered single- and double-metal hydroxide precipitates at the mineral/water interface: A multiple-scattering XAFS analysis. *Journal of Colloid and Interface Science* **2000**, *223*, 1-12.

(55) Bode, H.; Dehmelt, K.; Witte, J. Zur Kenntnis der Nickelhydroxidelektrode-I. Über das Nickel (II)-Hydroxidehydrat. *Electrochimica Acta* **1966**, *11*, 1079-1087.

(56) Mansour, A. N.; Melendres, C. A. Analysis of X-ray absorption spectra of some nickel oxycompounds using theoretical standards. *Journal of Physical Chemistry A* **1998**, *102*, 65-81.

(57) Manceau, A.; Drits, V.; Silvester, E.; Bartoli, C.; Lanson, B. Structural mechanism of Co^{2+} oxidation by the phyllomanganate buserite. *American Mineralogist* **1997**, *82*, 1150-1175.

(58) Manceau, A.; Schlegel, M.; Nagy, K. L.; Charlet, L. Evidence for the formation of trioctahedral clay upon sorption of Co^{2+} on quartz. *Journal of Colloid and Interface Science* **1999**, *220*, 181-197.

(59) Halbert, A. R.; Gebauer, K. A.; Wall, L. M. Prognosis of occupational chromate dermatitis. *Contact Dermatitis* **1992**, *27*, 214-219.

(60) Guo, Y. L.; Wang, B. J.; Yek, K. C.; Wang, J. C.; Kao, H. H.; Wang, M. T.; Shih, H. C.; Chen, C. J. Dermatoses in cement workers in Siouthern Taiwan. *Contact Dermatitis* **1999**, *40*, 1-7.

(61) Geier, J.; Schnuch, A. A comparison of contact allergies among construction and non-construction workers attending contact dermatitis clinics in Germany. Results of the IVDK from November 1989 until July 1993. *American Journal of Contact Dermatitis* **1995**, *6*, 86-94.

(62) Tokunaga, T. K.; Wan, J.; Hazen, T. C.; Schwartz, E.; Firestone, M. K.; Sutton, S. R.; Newville, M.; Olson, K. R.; Lanzirotti, A.; Rao, W. Distribution of chromium contamination and microbial activity in soil aggregates. *Journal of Environmental Quality* **2003**, *32* (2), 541-549.

(63) Peterson, M. L.; Brown Jr., G. E.; Parks, G. A.; Stein, C. L. Differential redox and sorption of Cr(III/VI) on natural silicate and oxide minerals: EXAFS and XANES results. *Geochimica et Cosmochimica Acta* **1997**, *61* (16), 3399-3412.

(64) Ross, D. S.; Hales, H. C.; Sea-McCarthy, G. C.; Lanzirotti, A. Sensitivity of soils manganese oxides: XANES spectroscopy may cause reduction. *Soil Science Society of America Journal* **2001**, *65*, 744-752.

CHAPTER 3

SPECTROSCOPIC INVESTIGATIONS OF NI SPECIATION IN HARDENED CEMENT PASTE

Abstract

Cement-based materials play an important role in multi-barrier concepts developed worldwide for the safe disposal of hazardous and radioactive wastes. Cement is used to condition and stabilize the waste materials and to construct the engineered barrier systems (container, backfill and liner materials) of repositories for radioactive waste. In this study, Ni uptake by hardened cement paste has been investigated with the aim of improving our understanding of the immobilization process of heavy metals in cement on the molecular level. X-ray-absorption spectroscopy (XAS) coupled with diffuse reflectance spectroscopy (DRS) techniques were used to determine the local environment of Ni in cement systems. The Ni-doped samples were prepared at two different water/cement ratios (0.4, 1.3) and hydration times (1 hour - 1 year) using a sulphate-resisting Portland cement. The metal loadings and the metal salts added to the system were varied (50 up to 5000 mg/kg; NO_3^-, SO_4^{2-}, Cl^-). The XAS study showed that for all investigated systems Ni(II) is predominantly immobilized in a layered double hydroxide (LDH) phase, which was corroborated by DRS measurements. Only a minor extent of Ni(II) precipitates as Ni-hydroxides (α-Ni(OH)$_2$ and β-Ni(OH)$_2$). This finding suggests that Ni-Al LDH, rather than Ni-hydroxides, is the solubility-limiting phase in the Ni-doped cement system.

3.1 Introduction

Assuring safe disposal and long-term storage of hazardous and radioactive wastes represents a primary environmental task of industrial societies. The long-term disposal of the hazardous waste is associated with landfilling of cement-stabilized waste (e.g. 1), whereas deep geological disposal is foreseen for some categories of cement-stabilized radioactive waste (2). For example, more than 90 wt% of the near-field material of the planned Swiss geological repository for intermediate-level waste consists of hardened cement paste (HCP) and cementitious backfill materials. The HCP is used to solidify the radioactive waste. For this reason, an understanding of the immobilization processes within a hydrating cement is essential to predict the long term fate of contaminants in the geosphere. From a chemical standpoint, HCP is a very heterogeneous material with discrete particles in the nano to micrometer size range. The material consists of mainly calcium (aluminum) silicate hydrates, portlandite (calcium hydroxide) and calcium aluminates. The immobilization potential of HCP originates from its selective binding properties for metal cations and anions (3). Thus, it appears that immobilization processes in cement systems are highly specific with respect to the mineral components and mechanisms involved.

Ni is among one important contaminant in waste materials resulting from a variety of industrial processes. For example, Ni radioisotopes associated with irradiated metallic components from nuclear power plants are present in cement-stabilized radioactive waste. In this case, these informations are of major importance for predicting the long-term fate of Ni in the cementitious matrix of a disposal site for hazardous or radioactive waste. In connection with the disposal of non-radioactive waste, molecular level information will allow more detailed assessment of the leachability of heavy metals, e.g. Ni, from landfills into the environment. Earlier experiments on the Ni uptake by blended and Portland cement indicated that, under highly alkaline conditions (pH>12.5), poorly crystalline $Ni(OH)_2$ (4) and Ni-Al layered double hydroxide (Ni-Al LDH) phases (5,6) may be formed. LDH-phases can be commonly expressed as $[M^{II}_{1-x}M^{III}_{x}(OH)_2]^{x+}(A^{n-})_{x/n} \cdot yH_2O$, where the M^{II} position can be at least partially filled with Ni, M^{III} position with Al and the A^{n-} with different anions such as CO_3^{2-}, NO_3^-, Cl^-, SO_4^{2-}. A natural occurring Ni-Al LDH mineral is Takovite, also named Eardleyit ($Ni_6Al_2(OH)_{16}CO_3 \cdot 4H_2O$).

The objective of the present study was to investigate Ni immobilization during cement hydration. The hydration process was started by adding Ni salt solution to the unhydrated cement. The sulphate-resisting Portland cement previously used by Scheidegger et al. (5,6) was also employed as starting material for the present study. Note that the experimental set-up differs from that used in the earlier studies (5,6) where the Ni uptake by hydrated cement was investigated.

A combination of wet chemistry experiments, X-ray-absorption spectroscopy (XAS) and diffuse reflectance spectroscopic (DRS) measurements was used to gain a molecular level understanding of the immobilization processes. The synergistic use of these techniques was expected to provide information on the solubility-limiting phase, chemical speciation and the structural coordination environment of Ni in hydrated cement.

3.2 Materials and methods

3.2.1 Sample preparation

The cement samples were prepared from a commercial sulphate-resisting Portland cement (CEM I 52.5 N HTS, Lafarge, France) used to condition radioactive waste in Switzerland. Ni-doped HCP was prepared by mixing different Ni salt solutions (NO_3^-, SO_4^{2-}, Cl^-) with unhydrated cement. The metal salts were dissolved in deionized water to obtain stock solutions with concentrations of 0.3, 0.03 and 0.003 mol/L (pH=4.5-5). The solutions were mixed with the unhydrated cement at two different water/cement (w/c) ratios (0.4, 1.3) using a standard procedure (7). The degree of hydration is enhanced with increasing w/c ratio (8). The final metal concentrations of the pastes were 50, 500 and 5000 mg/kg dry (Table 3.1). The cement pastes were filled into Plexiglas moulds, which were closed with polyethylene lids, and hydrated between 1 hour and ~1 year. For short hydration times up to 6 hours, the slurries were filtered (0.2 μm pore size) to separate the solid from the free water. The solid materials were washed with acetone for 15 minutes to stop the hydration process (8), filtered and dried in a glovebox under controlled N_2 atmosphere (CO_2, O_2<2 ppm, T=20±3 °C). The samples hydrated for longer time periods were stored in closed containers at 100% relative humidity. The cylinders were cut into several slices of ~1 cm thickness and dried in the glovebox. Some slices were crushed to obtain size fractions <100 μm using a tungsten/carbide pebble mill. The powder material was employed for extended X-ray absorption fine structure (EXAFS) and DRS measurements as well as for wet chemistry experiments. For EXAFS measurements the powder material was filled into Plexiglas holders and sealed with Kapton tape.

Table 3.1 Experimental parameters for the Ni-doped samples.

Sample name	Salt	Initial Ni concentration in solution (M)	w/c	Target metal concentration in HCP (mg/kg)	Hydration time
Ni_cem_1h	Ni(NO$_3$)$_2$	0.3	0.4	5000	1 hour
Ni_cem_6h	Ni(NO$_3$)$_2$	0.3	0.4	5000	6 hours
Ni_cem_3d	Ni(NO$_3$)$_2$	0.3	0.4	5000	3 days
Ni_cem_30d	Ni(NO$_3$)$_2$	0.3	0.4	5000	30 days
Ni_cem_150d	Ni(NO$_3$)$_2$	0.3	0.4	5000	150 days
Ni_cem_1y	Ni(NO$_3$)$_2$	0.3	0.4	5000	1 year
Ni_cem_500	Ni(NO$_3$)$_2$	0.03	0.4	500	30 days
Ni_cem_50	Ni(NO$_3$)$_2$	0.003	0.4	50	30 days
Ni_cem_w/c_1.3	Ni(NO$_3$)$_2$	0.3	1.3	5000	30 days
Ni_cem_Cl	NiCl$_2$	0.3	0.4	5000	30 days
Ni_cem_SO$_4$	Ni(SO$_4$)	0.3	0.4	5000	30 days

w/c = water/cement ratio

3.2.2 Wet chemistry experiments

The wet chemistry experiments were conducted using Ni-doped HCP samples reacted for 30 days with Ni loadings of 5000 mg/kg (Ni_cem_30d) and 500 mg/kg (Ni_cem_500, Table 3.1). Prior to use in the wet chemistry experiments the HCP material was crushed and sieved to collect the size fraction of <63 μm. The material was mixed with artificial cement pore water (ACW) at a solid/liquid (S/L) ratio of 5 g/L and shaken end-over-end for 1, 14, 30 and 60 days in the glovebox under controlled N$_2$ atmosphere. After equilibration solid and liquid phases were separated by centrifugation (60 min at 95000 g). Aliquots were withdrawn from the supernatant solution to determine the Ni concentration using inductively-coupled plasma mass spectroscopy (ICP-MS, detection limit=0.05 μg/L). Prior to the ICP-MS measurements the concentration of the main elements (Na, K, Ca) was determined using inductively-coupled plasma optical emission spectroscopy (ICP-OES) to take into account matrix effects.

The composition of ACW corresponds to the chemical composition of a solution in equilibrium with a freshly prepared HCP (8). Under these conditions the ACW is a (Na,K)OH rich fluid saturated with respect to portlandite and calcite (pH=13.3). The basic preparation and the chemical composition of ACW are described elsewhere (9,10).

3.2.3 EXAFS data collection and reduction

EXAFS spectra at the Ni K-edge were collected at the Swiss Norwegian Beam Line (SNBL) and at the Dutch Belgium Beamline (DUBBLE) at the European Synchrotron

Radiation Facility (ESRF) in Grenoble, France. Both beamlines are equipped with a Si(111) crystal monochromator. The measurements were collected at room temperature in transmission (ionization chambers) and in fluorescence mode (SNBL: Lytle detector; DUBBLE: 9 channel monolithic Ge-solid state detector). The monochromator angle was calibrated by assigning the energy of 8333 eV to the first inflection point of the K-absorption edge spectrum of Ni metal foil.

EXAFS data reduction was performed using the WinXAS 3.1 software package following standard procedures (11). The energy was converted to photoelectron wave vector units ($Å^{-1}$) by assigning the origin E_0 to the first inflection point of the absorption edge. Radial Structure Functions (RSF) were obtained by Fourier transforming the k^3-weighted $\chi(k)$ functions between 3.2 and 10.9 $Å^{-1}$ with a Bessel window function with a smoothing parameter of 4. Multishell fits were performed in real space across the range of the first two shells (ΔR=0.8-3.5 Å). Theoretical scattering paths for the fit were calculated using FEFF 8.2 (12) and the structure of β-Ni(OH)$_2$ as a reference. The amplitude reduction factor (S_0^2) was determined to be 0.85 from the experimental β-Ni(OH)$_2$ EXAFS spectrum (5). Errors on the structural parameters were estimated from the analysis of a series of reference compounds (β-Ni(OH)$_2$, α-Ni(OH)$_2$; see Table 3.2). Several reference spectra (β-Ni(OH)$_2$, α-Ni(OH)$_2$, synthetic Ni-Al LDH (Ni:Al, 2:1; Ni$_2$Al(OH)$_6$(CO$_3$)$_{1/2}$ (13)), Ni-phyllosilicate (14), neo-formed Ni-Al LDH formed from Ni-doped pyrophyllite (15)) were used in support of the identification of the Ni speciation in the cement matrix.

Wavelet Transform analysis (WT) of EXAFS spectra was used to complement the Fourier Transform. WT enables a 2D visualization of the Fourier Transform, with resolution in both k ($Å^{-1}$) and R space (Å) (16,17). WT allows to distinguish atoms, which are located at the same distance R but yield contributions at different k ranges. Recently, Muñoz et al. (17) applied a continuous Cauchy WT to simultaneously decompose the EXAFS spectra of geochemical and environmental materials in reciprocal and real space. In a similar manner, Funke et al. (16) applied the Morlet wavelet to resolve the short-range structure of Zn-Al LDH. The Morlet WT allows the optimization of the parameters η and σ. The frequency, η of the sine and cosine functions, determines how many oscillations of the sine wave are covered by a Gaussian envelope with the half-width σ=1. Funke et al. (16) demonstrated that the optimum resolution at a given distance (r_{opt}) of interest is achieved by $\eta_{opt} \cong 2r_{opt}$ for σ=1.

3.2.4 Diffuse Reflectance Spectroscopy (DRS)

DRS has been used in the past to study Ni and other transitional metals in catalysis (e.g. 18) and in environmental science (e.g. 19,20). The information provided by DRS is related to the local symmetry of the first coordination shell. For example, DRS is more sensitive to relative changes in the Ni-O distance than EXAFS. Scheinost et al. (19) showed the potential of DRS in identifying different Ni phases based on the energy position of the ν2 band. This band corresponds to the $^3A_{2g} \rightarrow {}^3T_{1g}$ transition of the crystal field.

In this study the DRS experiments were conducted using a Varian Cary 5 UV-vis-NIR spectrophotometer described elsewhere (e.g. 21). White reflectance standard $BaSO_4$ (Kodak) was used to record the base line. Processing of the spectra included subtraction of the base line and calculation of the Kubulka-Munk function. Blanks, i.e., samples prepared in the same way as the Ni-doped HCP samples but without metal addition, were used for spectral background subtraction. The spectra of reference compounds (synthetic Ni-Al LDH (Ni:Al, 2:1), commercial $Ni(NO_3)_2$ and $β-Ni(OH)_2$) were also recorded at the same time.

3.3 Results and discussion

3.3.1 Wet chemistry data

Wet chemistry experiments were carried out to quantify the Ni partition between the Ni-doped HCP and ACW, and compare the results with earlier measurements of the Ni sorption isotherm on HCP prepared from the same type of cement (10). Wet chemistry data allow first assessments of the binding mechanism of Ni in the cement matrix to be made. In this study the Ni partition between the Ni-doped HCP samples with Ni loadings of 5000 mg/kg dry HCP (Ni_cem_30d) and 500 mg/kg dry HCP (Ni_cem_500) was determined after 30 days equilibration of the suspensions. Preliminary kinetic tests over the time period of 60 days revealed that 7 days equilibration was the minimum time required to achieve constant Ni concentrations in solution (data not shown). Nevertheless, 30 days equilibration was chosen to allow direct comparison of the aqueous Ni concentrations in these systems with the concentration measurements reported by Wieland et al. (10).

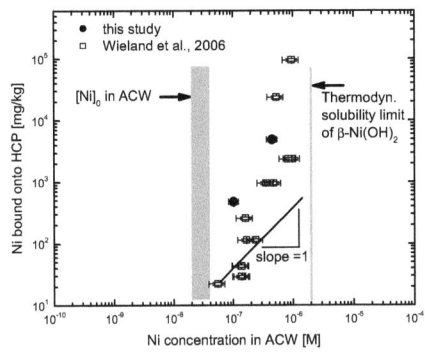

Fig. 3.1. The concentration of Ni taken up by HCP shown as function of the concentration of dissolved Ni in ACW (pH=13.3). Closed circles: 5000 and 500 mg/kg metal loading (this study); open squares: data from Wieland et al. (10). All samples were equilibrated for 30 days. The shaded area indicates the Ni concentration in ACW; the grey line indicates the solubility limit of β-Ni(OH)$_2$.

Fig. 3.1 shows the Ni concentration in solution upon equilibration of the Ni_cem_30d and Ni_cem_500 sample material in ACW together with the sorption isotherm data reported in Wieland et al. (10). The figure reveals that the trend to increasing aqueous Ni concentrations with increasing Ni loadings is comparable in both systems. Thus, similar Ni concentrations are determined regardless of the initial Ni source, that is whether Ni was released from the Ni doped HCP or added to solution and subsequently sorbed onto HCP. As discussed by Wieland et al. (10), the Ni concentration in ACW lies above the Ni concentration of pristine HCP equilibrated in ACW and below the solubility of β-Ni(OH)$_2$, indicating that the solubility-limiting phase is not β-Ni(OH)$_2$. It should be noted that for the experimental set-up used in the present study (adding a highly concentrated Ni solution to the alkaline cement system) and based on the available thermodynamic data (22 and therein), β-Ni(OH)$_2$ was expected to be the only Ni phase formed during cement hydration. The good agreement of the wet chemistry data obtained in this study and those reported by Wieland et al. (10) suggests that the same mechanism might be responsible for the Ni uptake by HCP regardless of the different methods used for the sample preparation. It is worth emphasizing, that, in contrast to the earlier studies (5,6,10) in the present study Ni was taken up by the cement matrix during the hydration process. Wieland et al. (10) concluded that the Ni immobilization by HCP cannot be interpreted in terms of an adsorption-type process. The authors rather suggested that the Ni concentration in solution is controlled by a solubility-limiting process due to the formation of a solid phase with varying composition (solid solution). This finding is further supported by the spectroscopic studies of Scheidegger et al. (5,6), which suggested that a Ni-Al LDH-type phase is the solubility-limiting phase.

3.3.2 Formation of a Ni-Al LDH phase

Fig. 3.2a shows the normalized, background-subtracted and k^3-weighted EXAFS spectra of HCP samples with Ni loadings of 5000 mg/kg and hydrated for different times together with relevant reference spectra (Table 3.1). For the following discussion, samples with hydration times longer than 30 days (Ni_cem_30d, Ni_cem_150d, Ni_cem_1y) will be considered.

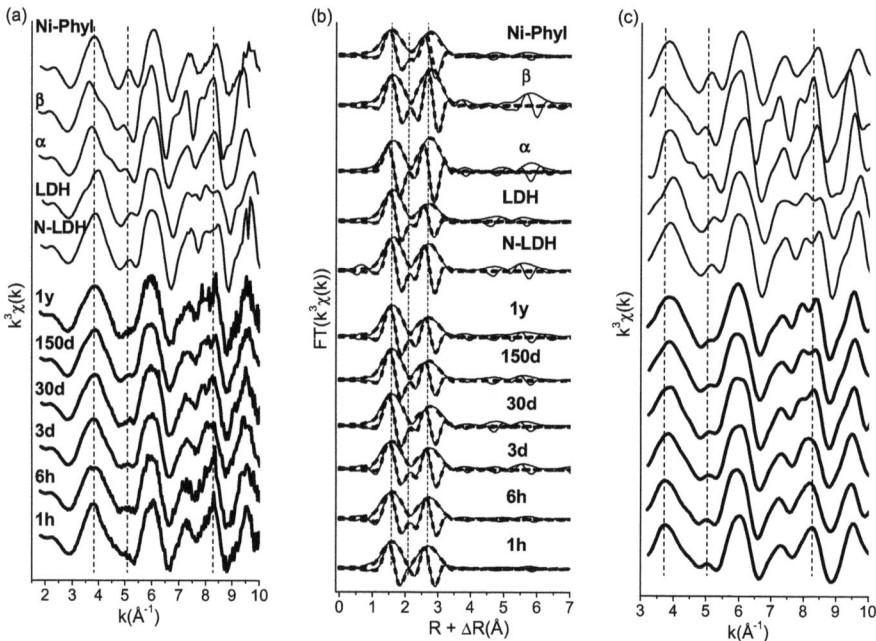

Fig. 3.2. Experimental spectra of Ni-reference compounds and Ni-doped HCP samples with 5000 mg/kg Ni concentration and hydrated between 1h and 1y a) k^3-weighted, normalized, background-subtracted EXAFS spectra, b) Experimental (solid line) and theoretical (dashed line) Fourier Transforms (modulus and imaginary parts) obtained from the EXAFS spectra presented in Fig. 3.2a, c) k^3-weighted EXAFS function for the Fourier-backtransform spectra obtained from Fig. 3.2b (range: R+ΔR=0.8-7 Å). Dashed lines indicate spectral features explained in detail in the text. N-LDH = neo-formed Ni-Al LDH (15), LDH = synthetic Ni-Al LDH (Ni:Al, 2:1), α = α-Ni(OH)$_2$, β = β-Ni(OH)$_2$, Ni-Phyl = Ni-phyllosilicate (14).

Fig. 3.2a reveals that the first oscillation at ~4 Å^{-1} of the experimental spectra is at a similar position as that of the neo-formed LDH phase, which forms upon Ni uptake by pyrophyllite (N-LDH) (15). The experimental spectra show a small feature at ~5 Å^{-1} on the left side of the oscillation at 6 Å^{-1}, which is well reproduced in both synthetic and neo-formed Ni-Al LDH spectra. The amplitude of this spectral feature is clearly damped compared to the Ni-phyllosilicate spectrum. The beat pattern at ~8 Å^{-1} of the Ni-doped HCP samples show a splitting of the oscillation. Scheinost and Sparks (23) demonstrated that the beat pattern at ~8 Å^{-1} suggests the presence of Ni-Al LDH. In fact, this beat pattern is observed in both Ni-Al LDH spectra, whereas the other reference compounds (α-Ni(OH)$_2$, β-Ni(OH)$_2$ and Ni-phyllosilicate) show an elongated upward oscillation ending in a sharp tip at ~8.5 Å. Thus, the presence of the beat pattern at ~8 Å^{-1} together with the observed spectral features at ~4 Å^{-1} and ~5 Å^{-1}, indicate that a Ni-Al LDH phase has formed in Ni-doped HCP samples.

The corresponding Fourier transforms (FT) of the k^3-weighted EXAFS spectra are shown in Fig. 3.2b. The amplitude of the first peak (R+ΔR=~1.6 Å) is similar for all samples and references. The imaginary part of the FT spectra of the Ni-doped HCP samples at R+ΔR=~2 Å show similarities with the α-Ni(OH)$_2$ and Ni-Al LDH reference compounds. The position of the second peak of the samples hydrated for 150 days (Ni_cem_150d) and for 1 year (Ni_cem_1y) is slightly shifted to lower values (R+ΔR=~2.71 Å) compared to the sample hydrated for 30 days (Ni_cem_30d) sample (R+ΔR=~2.78 Å). The latter distance is comparable to those determined for the β-Ni(OH)$_2$ and Ni-phyllosilicate reference compounds (R+ΔR=~2.75 Å). However, the shorter distance corresponds to those determined for α-Ni(OH)$_2$ or Ni-Al LDH, respectively (R+ΔR=~2.71 Å). The amplitude of the second peak in the experimental spectra of all Ni-doped HCP samples and the Ni-Al LDH reference compounds is clearly reduced compared to the other samples, supporting the presence of Ni-Al LDH in the cement samples.

The structural parameters derived from multi-shell analysis (ΔR=0.8-3.5 Å) are summarized in Table 3.2. The first coordination shell was fitted with Ni-O backscattering pairs. The second coordination shell was fitted solely using Ni-Ni pairs, because the discrimination of Ni-Ni and Ni-Al backscattering pairs in Ni-Al LDH is problematic (5,6). To be able to compare the coordination numbers of the Ni-Ni backscattering pairs (CN$_{Ni-Ni}$) of all samples and references, the Debye-Waller factor was fixed to 0.005 Å^2 as determined for the β-Ni(OH)$_2$. Data analysis reveals similar CN and interatomic distances (R) for all cement samples (Table 3.2). The first FT peak corresponds to an octahedral coordination of Ni with

~6 oxygen atoms at 2.04-2.05 Å. The CN_{Ni-Ni} (~3) are strongly reduced compared to α-Ni(OH)$_2$ and β-Ni(OH)$_2$. However, the CN_{Ni-Ni} of the HCP samples are similar to that determined for Ni-Al LDH. The CN_{Ni-Ni} is reduced as Ni is partly substituted by Al in Ni-Al LDH, resulting in a significant destructive interference between Ni and Al backscattering contributions and causing an amplitude cancellation of the Ni and Al shells (6,14). The CN_{Ni-Ni} in β-Ni(OH)$_2$ is close to 6 as expected from literature data (24), whereas the CN_{Ni-Ni} of α-Ni(OH)$_2$ is slightly lower (4.9). The difference in the CN is attributed to Ni vacancies in the brucite-like Ni(OH)$_2$ layer (25).

Although the CN_{Ni-Ni} of the Ni-doped HCP samples and Ni-Al LDH agree very well, the overall Ni-Ni distances of the HCP samples are longer (R_{Ni-Ni}=3.09-3.11 Å) compared to 3.06-3.07 Å (Ni-Al LDH). This finding suggests that, in addition to Ni-Al LDH phase other Ni-containing phases form. The longer R_{Ni-Ni} is attributed to the presence of β-Ni(OH)$_2$ impurities (R_{Ni-Ni}=3.13 Å). β-Ni(OH)$_2$ is expected to form because the Ni solution used for the sample preparation is strongly oversaturated with respect to Ni-hydroxide upon contact with cement (pH of the pore solution ~12.8-13.3).

Table 3.2. Structural information obtained from EXAFS Ni K-edge data analysis.

Samples	Ni-O			Ni-Ni			Ni-Si			ΔE$_0$ (eV)	Res%
	CN	R (Å)	σ² (Å²)	CN	R (Å)	σ² (Å²)	CN	R (Å)	σ² (Å²)		
References											
Ni-Phyllosilicate [a]	5.1	2.04	0.006	3.5	3.07	0.008[c]	3.7	3.26	0.008[c]	0.3	3.0
β-Ni(OH)$_2$	5.6	2.06	0.005	5.6	3.13	0.005				-0.6	3.0
α-Ni(OH)$_2$	5.2	2.03	0.005	4.9	3.09	0.005[d]				3.0	4.4
Ni-Al LDH (LDH)	6.0	2.05	0.006	2.5	3.06	0.005[d]				1.1	4.5
Neo-formed Ni-Al LDH [b] (N-LDH)	5.7	2.04	0.004	3.9	3.07	0.005[d]				0.3	3.8
Cement samples											
Ni_cem_1h	4.9	2.04	0.005	3.8	3.09	0.005[d]				-0.7	4.4
Ni_cem_6h	4.9	2.03	0.004	3.8	3.08	0.005[d]				-2.4	4.3
Ni_cem_3d	6.9	2.05	0.008	3.2	3.10	0.005[d]				0.0	7.6
Ni_cem_30d	7.3	2.05	0.007	3.0	3.11	0.005[d]				1.3	7.9
Ni_cem_150d	5.9	2.04	0.005	2.9	3.09	0.005[d]				-1.7	2.9
Ni_cem_1y	6.5	2.04	0.006	3.0	3.09	0.005[d]				-1.8	4.1
Ni_cem_500	5.3	2.05	0.005[d]	2.3	3.12	0.005[d]				0.0	9.0
Ni_cem_50	6.3	2.04	0.005[d]	1.8	3.17	0.005[d]				-0.3	10.8
Ni_cem_w/c_1.3	5.5	2.05	0.004	3.0	3.10	0.005[d]				-0.9	4.3
Ni_cem_Cl	5.0	2.04	0.004	2.9	3.10	0.005[d]				-0.5	3.5
Ni_cem_SO$_4$	6.4	2.05	0.007	2.7	3.08	0.005[d]				-0.7	4.0

[a] Dähn et al. 2002, [b] Scheidegger et al., 1997, [c] correlated parameters and [d] fix paramaters during fitting procedures.
CN, R, σ², ΔE$_0$ stand for interatomic distance, coordination number, Debye-Waller factor and inner potential correction.
Estimated errors: $R_{(Ni-O)}$ ±0.02 Å, $CN_{(Ni-O)}$ ±20%, $R_{(Ni-Ni)}$ ±0.02 Å, $CN_{(Ni-Ni)}$ ±20%
Res%: deviation between experimental data and fit given by the relative residual in percent.
N= number of data points, Y_{exp} and Y_{theo}: experimetnal and throretical data points, respectively.

$$\%Res = \frac{\sum_{i=1}^{N}|y_{exp}(i) - y_{theo}(i)|}{\sum_{i=1}^{N} y_{exp}(i)} * 100$$

The presence of Ni-Ni and Ni-Al backscattering contributions is further substantiated using WT analysis. Fig. 3.3 shows the WT of the sample hydrated for 1 year (Ni_cem_1y) (Fig. 3.3a) and the synthetic Ni-Al LDH (Fig. 3.3b), which were deduced using the optimized Morlet parameters $\eta=5.7$ and $\sigma=1$, and a k^3-weighted signal. Both samples show a maximum at R+ΔR=~1.6 Å and at k=~6 Å^{-1} with a contribution in k-space ranging between ~2 Å^{-1} and ~9 Å^{-1}. This corresponds to the first metal shell (Ni-O). The WT of both samples resolves two maxima at k=~6.6 Å^{-1} and k=~9.2 Å^{-1} for the peak at R+ΔR=~2.7 Å, respectively. Thus, based on the similarities of the WT, we infer that the second peak in the FT of the Ni-doped HCP sample and Ni-Al LDH consists of two identical contributions, resulting from Ni-Ni and Ni-Al backscattering pairs. This finding supports our hypothesis that Ni-Al LDH is formed in the cement samples.

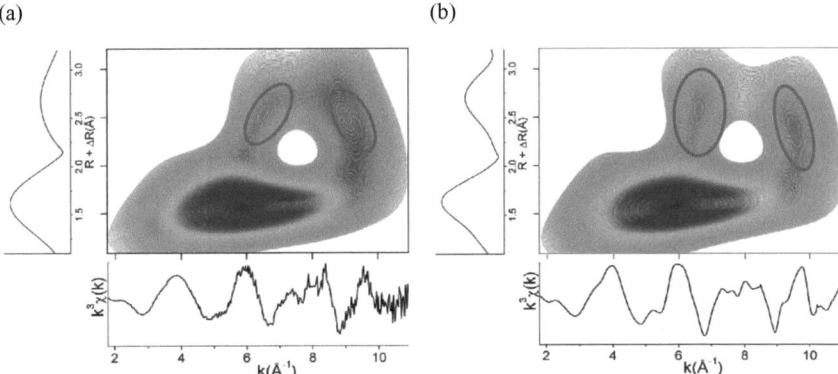

Fig. 3.3. Wavelet analysis of the first and second shell ($\eta=5.7$, $\sigma=1$, (16)) of a) Ni_cem_1 y sample (5000 mg/kg Ni loading, hydrated for 1 year), b) Ni-Al LDH synthetic reference compound (LDH, Ni:Al, 2:1). Circled areas indicate contributions of Ni-Ni and Ni-Al backscattering pairs. Note the slight difference in R between the experimental spectrum and the reference compound for the maxima at k=~6.6 Å^{-1} and ~9.2 Å^{-1}. This is due to the fact that the experimental spectrum is composed of a mixture of Ni-Al LDH and Ni-hydroxide phases, and not of a pure Ni-Al LDH, as for the reference compound. The Ni-hydroxides phases cause the R_{Ni-Ni} to shift to slightly longer distances R.

Independent evidence of the Ni-Al LDH formation in Ni-doped HCP samples was further obtained from DRS measurements. Fig. 3.4 shows the DRS measurements for the sample hydrated for 1 year (Ni_cem_1y) and the synthetic Ni-Al LDH. The Ni-Al LDH has a v2 band position at 15439 cm^{-1}, which is comparable to the position reported by Scheinost et al. (19)(15220 cm^{-1}-15430 cm^{-1}), and a small shoulder at 13480 cm^{-1}. The Ni_cem_1y sample shows a v2 band position at 15324 cm^{-1} and a small shoulder at ~13500 cm^{-1}, which is in good agreement with the reference compound. This indicates that predominantly Ni-Al LDH has formed in the cement sample.

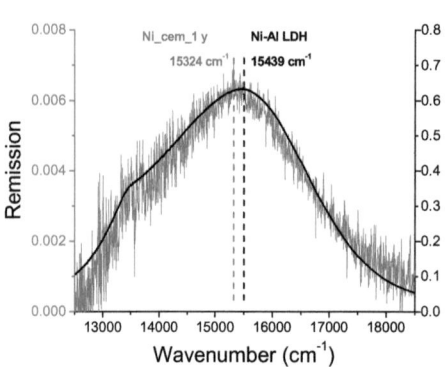

Fig. 3.4. DRS measurement of Ni_cem_1y sample (grey) (5000 mg/kg Ni loading, hydrated for 1 year) and Ni-Al LDH synthetic reference compound (black) (LDH, Ni:Al, 2:1). Note the different ordinates used for Ni_cem_1y (grey, left) and for Ni-Al LDH (black, right) and that the noise level of the Ni_cem_1y sample is higher compared to the reference due to lower Ni-Al LDH concentration.

3.3.3 Time dependency of the Ni-Al LDH formation

Samples with hydration times varying between 1 hour and 1year were investigated to asses the influence of the hydration time on Ni binding in hydrated cement. Fig. 3.2a shows the normalized, background-subtracted and k^3-weighted EXAFS spectra of Ni-doped HCP samples hydrated up to one year. In all experimental spectra the spectral features at ~4 Å^{-1}, ~5 Å^{-1} and ~8 Å^{-1} appear. The oscillation at ~4 Å^{-1} becomes broader with increasing hydration time. The position of this oscillation for the short hydration times (1 hour and 6 hours; Ni_cem_1h and Ni_cem_6h) is comparable to that of the Ni-hydroxides phases (Fig. 3.2a). In contrast, the position of this oscillation in the longer hydrated samples (3 days up to 1 year; Ni_cem_3d-Ni_cem_1y) corresponds to that of Ni-Al LDH. Furthermore, the right shoulder of this oscillation for the short hydration times (Ni_cem_1h and Ni_cem_6h) shows a small feature similar to α-Ni(OH)$_2$. The spectral feature detected at ~5 Å^{-1} shifts slightly to the right (dashed line) with increasing hydration time. The oscillation at ~8 Å^{-1} shows an elongated upward oscillation ending at ~8.5 Å for the Ni_cem_1h and Ni_cem_6h samples, similar to α-

Ni(OH)$_2$ (Fig. 3.2a). The EXAFS spectra of the Ni-doped HCP samples with hydration times longer than 3 days show a splitting of the oscillation at ~8 Å^{-1}. The three spectral features (~4 Å^{-1}, ~5 Å^{-1}, ~8 Å^{-1}) are better visualized in the Fourier-backtransform (FT^{-1}) shown in Fig. 3.2 (ΔR=0.8-7 Å). The broadening of the feature at ~4 Å^{-1}, the shift of the feature at ~5 Å^{-1} and the splitting of the oscillation at ~8 Å^{-1} indicate that the portion of Ni-Al LDH formed in HCP increases with increasing hydration time. At short hydration times, however, the slight shift of the feature at ~4 Å^{-1} and the shoulder of the feature at ~4 Å^{-1} suggest that the spectra of the Ni-doped HCP samples are very similar to those of the Ni-hydroxide phases.

The corresponding FT of the k^3-weighted EXAFS spectra (Fig. 3.2b) show no evident dependence on the hydration time. Data analysis reveals similar Ni-O and Ni-Ni distances for all Ni-doped HCP samples. However, the CN$_{Ni-Ni}$ decreases from 3.8 to 3.0 with increasing hydration time (Table 3.2). This decrease in the CN$_{Ni-Ni}$ is caused by destructive interference of Ni-Ni and Ni-Al backscattering pairs as discussed earlier. The finding of a significant reduction of CN$_{Ni-Ni}$ indicates substantial substitution of Ni by Al.

To test whether a mineral mixture would be consistent with the observed spectra, the EXAFS spectra of the Ni-doped HCP samples were fitted with linear combinations of α-Ni(OH)$_2$, β-Ni(OH)$_2$ and Ni-Al LDH. The amount of Ni-Al LDH in the Ni-doped HCP samples was found to increase with increasing hydration time (from ~40% to ~60%), whereas α-Ni(OH)$_2$ decreases accordingly (from ~35% to ~20%). Finally, the content of β-Ni(OH)$_2$ was found to be small (~20%) and remained constant with time.

3.3.4 Influence of other experimental parameters

Besides hydration time other experimental parameters, e.g. the initial Ni concentrations, the anions added to the system and the w/c ratio were expected to have a potential effect on the Ni speciation. Therefore, these parameters were varied to assess their influence on the Ni uptake by hydrating cement.

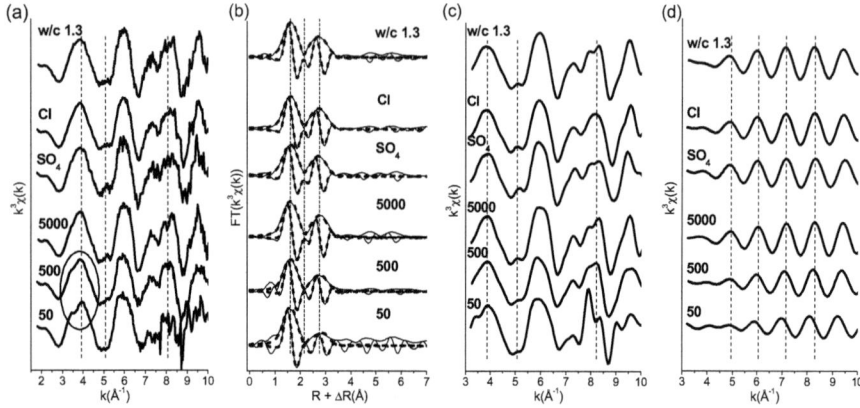

Fig. 3.5. Experimental spectra of Ni-doped HCP samples hydrated for 30 days at different concentrations (500, 50), using w/c 1.3 or different anions (Cl⁻, SO_4^{2-}). compared with Ni_cem_30d (NO_3^-, 5000 mg/kg, w/c 0.4) of a) k^3-weighted, normalized, background-subtracted EXAFS spectra, the circled area indicates features explained in the text, b) Experimental (solid line) and theoretical (dashed line) Fourier Transforms (modulus and imaginary parts) obtained from the EXAFS spectra presented in Fig. 3.5a, c) k^3-weighted EXAFS function for the Fourier-backtransform spectra obtained from Fig. 3.5b (range: R+ΔR=0.8-7 Å), d) k^3-weighted EXAFS function for the Fourier-backtransform spectra of the second shell obtained from Fig. 3.5b (range: R+ΔR=2.20-3.40 Å). Dashed lines indicate spectral features explained in detail in the text. 50 = Ni_cem_50, 500 = Ni_cem_500, 5000 = Ni_cem_30d, SO_4^{2-} = Ni_cem_SO₄, Cl⁻= Ni_cem_Cl, w/c 1.3 = Ni_cem_w/c_1.3.

Fig. 3.5 shows the normalized, background-subtracted and k^3-weighted EXAFS spectra (a), the FT (b), and the FT^{-1} (c) of Ni-doped HCP samples hydrated for 30 days. These samples were either doped with Ni(NO₃)₂ at different concentrations (50, 500, 5000 mg/kg; Ni_cem_50, Ni_cem_500, Ni_cem_30d) and different w/c ratios (0.4 and 1.3; Ni_cem_30d, Ni_cem_w/c_1.3), or the Ni salts used for the sample preparation were varied (NO_3^-, SO_4^{2-}, Cl⁻; Ni_cem_30d, Ni_cem_SO₄, Ni_cem_Cl) (Table 3.1). Fig. 3.5 a reveals that, for the Ni_cem_SO₄, Ni_cem_Cl and Ni_cem_w/c_1.3 samples, the oscillations at ~4 Å⁻¹, ~5 Å⁻¹ and ~8 Å⁻¹ are comparable to those previously observed for the Ni_cem_30d sample which was prepared at w/c=0.4 to obtain 5000 mg/kg Ni loading using Ni(NO₃)₂. Further, the derived FT of the k^3-weighted EXAFS spectra of these samples show no noticeable difference to the Ni_cem_30d sample (Fig. 3.5b). To better visualize the contributions resulting from the second shell, a FT^{-1} was performed in the range between 2.20-3.40 Å (Fig. 3.5d). The contribution of the second shell in the Ni_cem_SO$_4$, Ni_cem_Cl and Ni_cem_w/c_1.3 samples are very similar to that in the Ni_cem_30d sample. Thus, neither the w/c ratios nor the type of anions used for the preparation have any noticeable influence on the EXAFS spectra.

However, some changes in the EXAFS spectra appear at decreasing Ni concentration. The oscillation at ~4Å^{-1} shows a splitting, which becomes more pronounced in the sample with Ni loading of 50 mg/kg (Ni_cem_50). The oscillation at ~5Å^{-1} as well as the feature between ~7 Å- ~8 Å^{-1} diminish with decreasing Ni loading. The described spectral features (~4 Å^{-1}, ~5 Å^{-1} and ~8 Å^{-1}) are better resolved in the FT^{-1} (Fig. 3.5c). The second peak in the FT of the samples with Ni loadings of 500 mg/kg (Ni_cem_500) and 50 mg/kg (Ni_cem_50) show a shift to the right and a broadening with decreasing Ni concentration compared to the Ni_cem_30d sample. The FT^{-1} between 2.20-3.40 Å reveal a frequency shift to the left with decreasing Ni concentrations for the Ni_cem_500 and Ni_cem_50 samples indicating significant changes in the backscattering contributions (Fig. 3.5d).

Structural parameters derived from multi-shell analysis (Table 3.2) show similar CN_{Ni-O}, CN_{Ni-Ni} and $R_{Ni-O;Ni-Ni}$ for all experimental spectra, which are further comparable to structural parameters determined for the Ni_cem_30d sample. The only exception is the Ni_cem_50 sample, for which the multi-shell fit approach used throughout this study failed as it resulted in Ni-Ni distances (3.17 Å) longer than any reported distances for Ni compounds (< 3.13 Å). At the present time the situation regarding the Ni speciation at the lowest loading is unclear.

3.3.5 Controlling uptake mechanism of Ni in cement

From the present study it appears that the formation of the different Ni phases in the Ni-doped HCP samples is controlled by both kinetic and thermodynamic constraints. The EXAFS results show that the formation of Ni-Al LDH, which immediately starts upon the addition of a Ni solution to unhydrated cement, is a fast process. Nevertheless, the study further shows that, in addition to Ni-Al LDH, α/β Ni(OH)$_2$ form in the initial phase of the hydration process as the system is strongly oversaturated with respect to Ni-hydroxides. The wet chemistry data, however, suggest that β-Ni(OH)$_2$ is not the thermodynamically most stable phase in the systems. The portion of Ni-Al LDH formed increases with increasing hydration time, indicating that Ni-Al LDH is the thermodynamically most stable Ni-containing phase in the cement matrix. The content of β-Ni(OH)$_2$ was found to be small and remained constant with time, suggesting that the transformation of β-Ni(OH)$_2$ to α-Ni(OH)$_2$ or Ni-Al LDH, respectively, is kinetically hindered. By contrast, α-Ni(OH)$_2$ is transformed into Ni-Al LDH with increasing hydration time as revealed from a decrease of the amount of initially formed α-Ni(OH)$_2$ with time. Note that an enhanced stability of Ni-Al LDH over α-Ni(OH)$_2$ in Al-containing hyperalkaline solution was already indicated in the studies of Wang et al. (26). The results of the present study further corroborate the findings of Scheidegger et al. (5,6). These authors reported the formation of predominantly Ni-Al LDH and minor quantities of β-Ni(OH)$_2$ when Ni was sorb onto hydrated cement and allowed to react for 150 days. The results from this study and the earlier work of Scheidegger et al. (5,6) clearly show that both modes of Ni immobilization, i.e. during cement hydration and due to sorption onto hydrated cement, lead to identical Ni speciations.

The formation of Ni-Al LDH requires that an Al source is available over the entire period of the hydration process. Lothenbach and Wieland (8) recently investigated the hydration process of the cement used in the present study. The Al-containing clinker minerals, which slowly dissolve during hydration, are aluminate and ferrite. The most important Al-containing minerals of the modeled hydrate assemblage are ettringite (Ca$_6$Al$_2$(SO$_4$)$_3$(OH)$_{12}$·26H$_2$O; ~8 wt%), calcium-monocarboaluminate (3CaO·Al$_2$O$_3$·CaCO$_3$·11H$_2$O; ~8 wt%) and hydrotalcite (LDH, [Mg$_{1-x}$Al$_x$(OH)$_2$]$^{x+}$(A^{n-})$_{x/n}$·yH$_2$O; ~1.5 wt%). The formation of ettringite is a fast process, which stops after 24 h when the sulphate source (gypsum) is exhausted. After that the formation of calcium-monocarboaluminate and hydrotalcite starts, and the portion of the hydrates in the cement matrix slowly increases with time (8). In this study the formation of Ni-Al LDH was observed in the first hours of the hydration process, indicating competition between ettringite and Ni-Al LDH for Al. Nevertheless, a significant influence of the competitive reactions on the hydration process can be excluded due to the small amount of Ni-Al LDH formed at the given initial Ni concentrations.

3.4 References

(1)   Schmidt, M.; Beckefeld, P.; Götz, R.; Kamsties, S.; Kretz, C.; Molitor, N.; Neck, U.; Vogel, P. *Reststoff-und Abfallverfestigung. Immobilisierung von Schadstoffen-Recycling-Verbesserung der Deponiefähigkeit*; Expert Verlag: Renningen-Malmheim, **1995**.

(2) Chapman, N.; McCombie, C. *Principles and standards for the disposal of long-lived radioactive wastes*; 1 ed.; Elsevier Science, Ltd.: Oxford, **2003**.

(3) Glasser, F. P. Chemistry of cement-solidified waste forms. . In: *Chemistry and microstructure of solidified waste forms*; Spence, R. D., Ed.; Lewis Publishers: Boca Raton, **1993**.

(4) Atkins, M.; Glasser, F. P.; Moroni, L. P.; Jack, J. J. Thermodynamic modelling of blended cements at elevated temperatures (50°C to 90°C). *DoE/UMIP/PR/94.011* **1994**.

(5) Scheidegger, A. M.; Wieland, E.; Dähn, R.; Spieler, P. Spectroscopic evidence for the formation of layered Ni-Al double hydroxides in cement. *Environmental Science and Technology* **2000**, *34*, 4545-4548.

(6) Scheidegger, A. M.; Wieland, E.; Scheinost, A. C.; Dähn, R.; Tits, J.; Spieler, P. Ni phases formed in cement and cement systems under highly alkaline conditions: An XAFS study. *Journal of Synchrotron Radiation* **2001**, *8*, 916-918.

(7) Döhring, L.; Görlich, W.; Rüttener, S.; Schwerzmann, R. Herstellung von homogenen Zementsteinen mit hoher hydraulischer Permeabilität. *Nagra NIB 94-29* **1994**.

(8) Lothenbach, B.; Wieland, E. A thermodynamic approach to the hydration of sulphate-resisting portland cement. *Waste Management* **2006**, *26*, 706-719.

(9) Wieland, E.; Tits, J.; Spieler, P.; Dobler, J. P. Interaction of Eu(III) and Th(IV) with sulphate-resisting portland cement. *Materials Research Society Symposium Proceedings* **1998**, *506*, 573-578.

(10) Wieland, E.; Tits, J.; Ulrich, A.; Bradbury, M. H. Experimental evidence for solubility limitation of the aqueous Ni(II) concentration and isotopic exchange of ^{63}Ni in cementitious systems. *Radiochimica Acta* **2006**, *94*, 29-36.

(11) Ressler, T. WinXAS: A program for X-ray absorption spectroscopy data analysis under MS-Windows. *Journal of Synchrotron Radiation* **1998**, *5* (2), 118-122.

(12) Rehr, J. J.; Mustre de Leon, J.; Zabinsky, S. I.; Albers, R. C. Theoretical X-ray absorption fine structure standards. *Journal of the American Chemical Society* **1991**, *113*, 5135-5140.

(13) Johnson, C. A.; Glasser, F. P. Hydrotalcite-like minerals ($M_2Al(OH)_6(CO_3)_{0.5} \cdot XH_2O$), where M= Mg, Zn, Co, Ni) in the environment: synthesis, characterization and thermodynamic stability. *Clays and Clay Minerals* **2003**, *51*, 1-8.

(14) Dähn, R.; Scheidegger, A. M.; Manceau, A.; Schlegel, M. L.; Baeyens, B.; Bradbury, M. H.; Morales, M. Neoformation of Ni phyllosilicate upon Ni uptake on montmorillonite: A kinetics study by powder and polarized extended X-ray absorption fine structure spectroscopy. *Geochimica et Cosmochimica Acta* **2002**, *66* (13), 2335-2347.

(15) Scheidegger, A. M.; Lamble, G. M.; Sparks, D. L. The kinetics of nickel sorption on phyrophyllite as monitored by X-ray absorption fine structure (XAFS) spectroscopy. *Journal de Physique IV France* **1997**, *7* (C2), 773-775.

(16) Funke, H.; Scheinost, A. C.; Chukalina, M. Wavelet analysis of extended X-ray absorption fine structure data. *Physical Review B* **2005**, *71*, 094110-094111-094117.

(17) Muñoz, M.; Argoul, P.; Farges, F. Continuous Couchy wavelet transform analyses of EXAFS spectra: A qualitative approach. *American Mineralogist* **2003**, *88*, 694-700.

(18) Bensalem, A.; Weckhuysen, B. M.; Schoonheydt, R. A. In situ diffuse reflectance spectroscopy of supported chromium oxide catalysts: Kinetics of the reduction process with carbon monoxide. *Journal of Physical Chemistry B* **1997**, *101*, 2824-2829.

(19) Scheinost, A. C.; Ford, R. G.; Sparks, D. L. The role of Al in the formatin of secondary Ni precipitates on pyrophyllite, gibbsite, talc, and amorphous silica: A DRS study. *Geochimica et Cosmochimica Acta* **1999**, *63*, 3193-3203.

(20) Scheckel, K. G.; Sparks, D. L. Kinetics of the formation and dissolution of Ni precipitates in a gibbsite/amorphous silica mixture. *Journal of Colloid and Interface Science* **2000**, *229*, 222-229.

(21) Verberckmoes, A. A.; Uytterhoeven, M. G.; Schoonheydt, R. A. Framework and extraframework Co^{2+} in CoAPO-5 by diffuse reflectance spectroscopy. *Zeolites* **1997**, *19*, 180-189.

(22) Hummel, W.; Curti, E. Nickel aqueous speciation and solubility at ambient conditions: a thermodynamic elegy. *Monatshefte für Chemie* **2003**, *134*, 941-973.

(23) Scheinost, A. C.; Sparks, D. L. Formation of layered single- and double-metal hydroxide precipitates at the mineral/water interface: A multiple-scattering XAFS analysis. *Journal of Colloid and Interface Science* **2000**, *223*, 1-12.

(24) Mansour, A. N.; Melendres, C. A. Analysis of X-ray absorption spectra of some nickel oxycompounds using theoretical standards. *Journal of Physical Chemistry A* **1998**, *102*, 65-81.

(25) Bode, H.; Dehmelt, K.; Witte, J. Zur Kenntnis der Nickelhydroxidelektrode-I. Über das Nickel (II)-Hydroxidehydrat. *Electrochimica Acta* **1966**, *11*, 1079-1087.

(26) Wang, C. Y.; Zhong, S.; Konstantinov, K.; Walter, G.; Liu, H. K. Structural study of Al-substituted nickel hydroxide. *Solid State Ionics* **2002**, *148*, 503-508.

CHAPTER 4

MICRO-SCALE INVESTIGATIONS OF NI UPTAKE BY CEMENT USING A COMBINATION OF SCANNING ELECTRON MICROSCOPY AND SYNCHROTRON-BASED TECHNIQUES

Abstract

Cement is used to condition waste materials and for the construction and backfilling of repositories for low and intermediate level radioactive waste. In this study, Ni uptake by hardened cement paste has been investigated with the aim of improving our understanding of the immobilization process of Ni(II) in cement on the micro-scale. Information on the cement microstructure, Ni distribution, Ni concentration and speciation of the Ni phases formed in the cement system and their association with specific cement minerals has been gained by using scanning electron microscopy (SEM) and synchrotron-based µ-X-ray fluorescence (µ-XRF) and µ-X-ray absorption spectroscopy (µ-XAS). The Ni-doped samples were prepared at a water/cement ratio of 0.4 using a sulphate-resisting Portland cement and hydrated for 30 days. The metal loadings of the system were varied from 50 up to 5000 mg/kg. SEM investigations show that for all metal loadings the Ni phases form rims around inner-calcium-silicate-hydrates, suggesting a direct association with this cement phase. The µ-XAS measurements further reveal that a mixture of Ni phases form at Ni enriched regions. Data analysis indicates that Ni(II) is predominantly immobilized in a layered double hydroxide-type phase (Ni-Al LDH) and only to a minor extent precipitates as Ni-hydroxides (α-Ni(OH)$_2$ and β-Ni(OH)$_2$). At 50 mg/kg Ni loading, however, the µ-XAS measurements suggest the presence of an additional Ni species. In the latter system Ni-Al LDH is found in Ni-rich regions, whereas at Ni-poor regions an unknown species is formed.

4.1 Introduction

Cement-based materials have been used worldwide for stabilizing hazardous and radioactive wastes in order to prevent or lower the mobility of contaminants in the environment that are present in the waste matrices. The long-term disposal of cement-stabilized hazardous waste is associated with landfilling of these waste forms (e.g. 1), whereas deep geological disposal is foreseen for some categories of cement-stabilized radioactive waste (2).

Thus, a molecular-level understanding of the processes governing the immobilization of heavy metals in hydrating cement is essential for long-term predictions of the environmental impact of cement-stabilized waste forms. From a chemical standpoint, hardened cement paste (HCP) is a very heterogeneous material with discrete particles in the nano- to micrometer size range. The material consists of mainly calcium (aluminum) silicate hydrates (C-(A)-S-H), portlandite (calcium hydroxide), calcium aluminates and calcium ferrites. The immobilization potential of HCP originates from its selective binding properties for metal cations and anions (e.g. (3)). Therefore, it appears that immobilization processes in cement systems are highly specific with respect to the mineral components and mechanisms involved.

Ni is an important contaminant in waste materials, generated in various industrial processes, such as electricity production in nuclear power plants. In the past, only few X-ray absorption spectroscopic (XAS) studies have been reported on the Ni uptake by cement. Investigations of Ni sorption onto hydrated Portland cement revealed that a mixture of $Ni(OH)_2$ and Ni-Al layered double hydroxide (Ni-Al LDH, $[M^{II}_{1-x}M^{III}_{x}(OH)_2]^{x+}(A^{n-})_{x/n} \cdot yH_2O$) phases form in these systems (4). In a recent study, Vespa et al. (5) investigated the Ni speciation during cement hydration using XAS on powdered Ni-doped cement samples (bulk-XAS). The varied experimental parameters were hydration time, anions added to the system, water/cement (w/c) ratio and Ni concentration. The study revealed the predominant formation of Ni-Al LDH in the hydrating cement system (500 and 5000 mg/kg Ni loadings) with small amounts of Ni-hydroxide phases (α-$Ni(OH)_2$ and β-$Ni(OH)_2$). The study further showed that the amount of Ni-Al LDH increases as a function of hydration time, whereas α-$Ni(OH)_2$ decreases and β-$Ni(OH)_2$ remains constant. These findings were independent of the anions added to the system or the w/c ratio. The only exception was found in the sample with the lowest Ni loading (50 mg/kg), where the Ni speciation was not dominated by the presence of Ni-Al LDH and Ni-hydroxides.

The goal of this study was to investigate the Ni uptake by HCP on the micro-scale and to further investigate the influence of the inherent heterogeneity of the cement matrix on the Ni speciation. To achieve this goal scanning electron microscopy (SEM), µ-synchrotron-X-ray fluorescence (µ-XRF) and µ-XAS were combined. SEM allows spatially-resolved information on the microstructure of the clinker and hydrated cement phases as well as associations of the Ni phases with specific cement minerals to be gained. Additionally, µ-XRF allows larger areas to be investigated and is therefore well suited to gain an overview of the Ni distribution. To determine the chemical speciation of Ni at selected regions of the cement matrix µ-XAS was employed. The knowledge acquired through this study will allow the development of a molecular-level understanding of the Ni uptake processes in cement. Moreover, the comparison between the micro-scale investigation and the bulk-XAS measurements (5) will allow identification of the relevant Ni species formed in the cement matrix.

4.2 Materials and methods
4.2.1 Sample preparation

The cement samples were prepared from a commercial sulphate-resisting Portland cement (CEM I 52.5 N HTS, Lafarge, France) used to condition radioactive waste in Switzerland. Ni-doped HCP was prepared by mixing a $Ni(NO_3)_2$ solution with unhydrated cement. The metal salt was dissolved in deionized water to obtain three stock solutions with concentrations of 0.003, 0.03 and 0.3 mol/L (pH=4.5-5). The solutions were mixed with the unhydrated cement at a w/c ratio of 0.4, using a standard procedure (5). The final metal concentrations of the pastes were 50 (Ni_cem_50), 500 (Ni_cem_500) and 5000 (Ni_cem_5000) mg/kg dry HCP. The cement pastes were poured into plexiglass moulds, which were closed with polyethylene lids, and hydrated for 30 days. The samples were stored in closed containers at room temperature at 100% relative humidity. Upon hydration the cylinders were cut into several slices of ~1 cm thickness and dried in the glovebox in a dry N_2 atmosphere. Some slices were impregnated and polished for the preparation of thin sections, which were employed for SEM and µ-XRF/µ-XAS measurements.

4.2.2 Scanning electron microscopic investigations

The SEM experiments were conducted at the Laboratory for Construction Materials (IMX), Ecole Polytechnique Fédéral de Lausanne (EPFL), using a FEI Quanta 200 microscope and at the Laboratory for Materials Behaviour (LWV), Paul Scherrer Institute (PSI), using a Zeiss DSM 962 microscope. The FEI microscope was operated at an accelerating voltage of 15 kV and a beam current of 100 µA, whereas the Zeiss was operated at an accelerating voltage of 20 kV and a beam current of 76 µA. The FEI microscope is equipped with a solid state detector, whereas the Zeiss is equipped with a photomultiplier for backscattering electron (BSE) images. Both microscopes are equipped with a Si(Li)-detector for energy dispersive micro-analysis (EDS). The sample-volume probed was ~1 µm^3.

BSE generates a specific phase contrast thanks to which minerals can be identified according to their brightness in the image. The minerals with the greatest average atomic number show the brightest contrast, whereas those with the lower atomic number show darker contrast. This allows the mineral components in the microstructure, both of unreacted clinker and hydrated cement phases, to be discriminated on the basis of their grey level. As a consequence, phases containing high concentrations of light elements such as calcium, hydrogen and oxygen (i.e. C-S-H) will be darker than other phases predominantly consisting of heavier elements such as Ni, Fe or oxides such as Ca_2SiO_4, Ca_3SiO_5.

The BSE images and elemental distribution maps, using an EDS system, were performed on the same polished thin sections used for the micro-spectroscopic investigations.

4.2.3 µ-XRF/µ-XAS data collection and reduction

µ-XRF/µ-XAS data were collected on beamline 10.3.2 ALS/USA (6). The monochromator angle was calibrated by assigning the energy of 8333 eV to the first inflection point of the K-edge absorption spectrum of Ni metal foil. The µ-XRF/µ-XAS measurements were collected at room temperature in fluorescence mode using a 7-element Ge-solid state detector with a beam size of ~5x5 µm^2. The µ-XRF maps were obtained by scanning the sample under the monochromatic beam at the energy of 10000 eV with a pixel size of 5x5 µm^2.

Data reduction was performed by standard procedures (see Supporting Information) using the WinXAS 3.1 software package (7). Several reference spectra (β-Ni(OH)$_2$, α-Ni(OH)$_2$, synthetic Ni-Al LDH (Ni:Al, 2:1; $Ni_2Al(OH)_6(CO_3)_{1/2}$ (8), Ni-phyllosilicate, neo-formed Ni-Al LDH formed from Ni-doped pyrophyllite), collected at the SNBL/ESRF, France, were used to identify the Ni species in the cement matrix (5).

4.3 Results and discussion

4.3.1 Distribution and speciation of the Ni phases

BSE-imaging allows minerals of different compositions to be identified. Fig. 4.1a-c show selected BSE images of the Ni-doped HCP sample at a w/c=0.4, hydrated for 30 days and with a metal concentration of 5000 mg/kg. In the first BSE image (Fig. 4.1a) the non-hydrated clinker mineral phases, belite (Ca_2SiO_4) and alite (Ca_3SiO_5) can be identified. Belite and alite are the major clinker phases in cement (9) with particles ranging in size between a few microns and tens of microns. During hydration, which starts upon addition of water to cement, alite and belite decompose to form calcium silicate hydrates (C-S-H) and calcium hydroxide (portlandite). C-S-H is the most abundant hydrated cement phase and confers most of its properties to the HCP. It precipitates under two main forms: inner-C-S-H and outer-C-S-H. The inner-C-S-H gradually fills up the space originally occupied by alite grains. After some time of hydration, most of the alite grains are surrounded by a rim of inner-C-S-H (white frame in Fig. 4.1a) varying in thickness. In some cases alite was found to be completely reacted and replaced by inner-C-S-H (black frame in Fig. 4.1a). On the other hand, the outer-C-S-H (darker grey area indicated in Fig. 4.1a and c) fills the porous space of the hydrated cement, originally occupied by water. The outer-C-S-H is generally less dense than the inner-C-S-H and finely intermixed with portlandite and several other minor hydrated phases, such as ettringite ($Ca_6Al_2(SO_4)_3(OH)_{12}\cdot 26H_2O$). The bright rims around inner-C-S-H phases found in Ni-doped HCP (red dashed area in Fig. 4.1c) were not observed in the non-doped HCP. The elemental distribution maps displayed in Fig. 4.1d-i show that Ni is accumulated in these bright rims. The Ni rim shown in Fig. 4.1 has a thickness of ~5 µm and a diameter of ~20–50 µm. This is an unusually large rim, which has been rarely observed in the investigated cement matrix. In general, rims of a few hundred nanometers up to a few microns thickness and up to ~10 µm in diameter have been observed, as shown in Fig. 4.2a-i. Although Fig. 4.1 and Fig. 4.2 show rims with different sizes, they display common features. The BSE images (Fig. 4.1a-c, Fig. 4.2a-c) show that the bright Ni-rich rims form around inner-C-S-H. The elemental distribution maps (Fig. 4.1d-i, Fig. 4.2d-i) reveal that Ni anti-correlates with Ca and Si, whereas correlation with Al and O is found. The Ca-Si anti-correlation is less obvious in Fig. 4.2 due to the large amount of small alite grains totally reacted to inner-C-S-H, which are surrounded by a relatively thin Ni rim (Fig. 4.2c red circle). Furthermore, in between these grains, outer-C-S-H has precipitated. The Ni-Al correlation is also less evident due to high Al concentrations. The significant correlation of Al with Fe, and to a lesser extent with Ca, visible at this particular spot (Fig. 4.2), suggests the presence of ferrite (4CaO·Al$_2$O$_3$·Fe$_2$O$_3$) and/or hydrated products of ferrite. No correlation of Ni with Fe was observed.

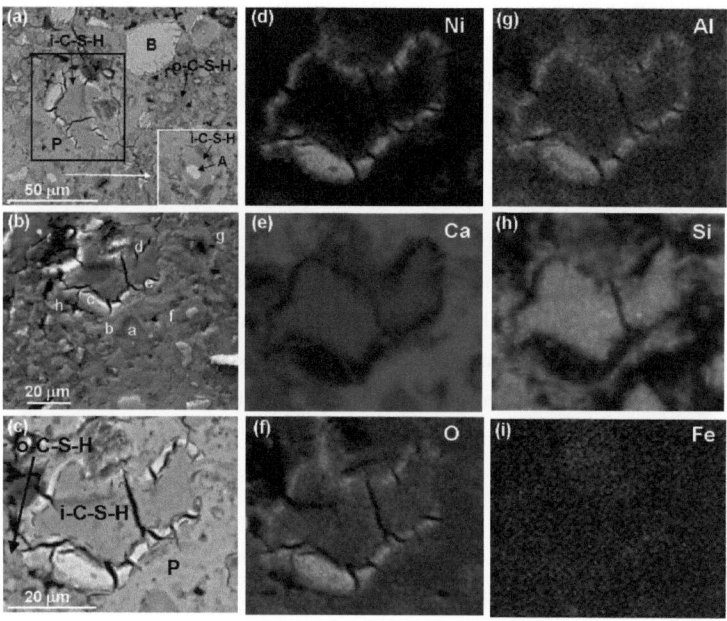

Fig. 4.1. BSE images (a-c) and SEM/EDS elemental distribution maps (d-i) of a Ni-rich region from the Ni(NO$_3$)$_2$-doped HCP sample with a water/cement ratio of 0.4 and a final metal concentration of 5000 mg/kg hydrated for 30 days. The white frame in (a) shows the enlarged region indicated by the arrow. In the enlarged picture an alite crystal is observed with the rim of inner-C-S-H. The black frame in (a) represents the 90° anticlockwise flipped image of the two BSE images observed in (b) and (c), representing the Ni-rich region, and the SEM/EDS maps (d-i). Note that in the SEM/EDS maps (d) to (i) brighter colors represent higher concentrations, whereas darker colors represent lower concentrations. The lower case letters a to h in (b) indicate the regions of the SEM/EDS single spot analyses. The red dashed area in (c) represents the Ni-rich region. Notations: B= belite (Ca$_2$SiO$_4$), A= alite (Ca$_3$SiO$_5$) P= portlandite (Ca(OH)$_2$), i-C-S-H= inner-calcium silicate hydrate, o-C-S-H= outer-C-S-H.

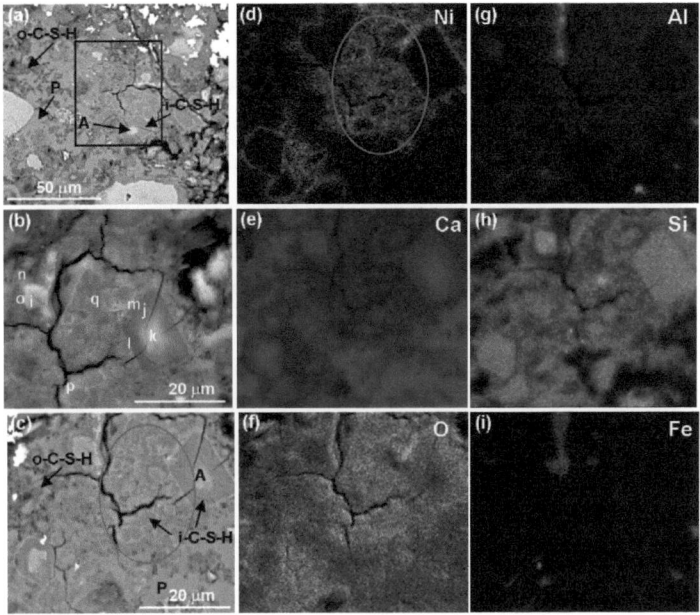

Fig. 4.2. BSE images (a-c) and SEM/EDS elemental distribution maps (d-i) of a further Ni-rich region with relatively fine rims (1-2 μm) of the $Ni(NO_3)_2$-doped HCP sample with a water/cement ratio of 0.4 and a final metal concentration of 5000 mg/kg hydrated for 30 days. The black frame in (a) represents the 90° anticlockwise flipped image of the other two BSE images observed in (b) and (c), representing the Ni-rich area, and the SEM/EDS maps (d-i). Note that in the SEM/EDS maps (d) to (i) brighter colors represent higher concentrations, whereas darker colors represent lower concentrations. The lower case letters i to p in (b) indicate the regions of the SEM/EDS single spot analyses. Notations: A= alite (Ca_3SiO_5) P= portlandite ($Ca(OH)_2$), i-C-S-H= inner-calcium silicate hydrate, o-C-H-S= outer-C-S-H. Red circle in (c) and (d) indicate features explained in the text.

The EDS analyses of selected spots on the Ni rims confirm the correlation between Ni and Al. The Al concentrations vary between ~1-3 wt%, whereas the Ni concentrations range between ~2-13 wt% for thinner (Table S4.1, analyses f-h and l-q, Supporting Information, indicated in Fig. 4.1b and Fig. 4.2b) and ~22-28 wt% for thicker rims (Table S4.1, analyses c-e indicated in Fig. 4.1b). The observed Ni:Al ratio is consistent with data published recently by Peltier et al. (10). Further EDS analyses of selected regions show that Ni is neither present in alite (Table S4.1, analyses i and k indicated in Fig. 4.2b) nor in inner-C-S-H or portlandite

(Table S4.1, analyses a, b indicated in Fig. 4.1b, and analysis j indicated in Fig. 4.2b). However, the EDS analyses show that Ca (~10-30 wt%) and Si (~2-10 wt%) are present in several percentages in the Ni-rich rims (Table S4.1, analyses c-h and l-q indicated in Fig. 4.1b and Fig. 4.2b). This seems to be in contrast with the elemental distribution maps, which revealed an anti-correlation of Ni with Ca and Si. Nevertheless, one must bear in mind that distribution maps are relative measurements based on the elemental concentration rather than absolute measurements. Finally, the EDS analyses indicate that further elements, which are present in trace concentrations in the cement matrix (Fe, S, Sr, Mg, Na, K, Cr, Cl, Table S4.1) are completely absent in the Ni rims.

A tentative explanation for the formation of Ni-rich rims around alite grains can be given by considering some specific features of this mineral. Firstly, alite dissolves much faster than belite, thus acting as a reactive zone in the cement matrix. Secondly, traces of Mg and Al are predominantly associated with alite (Mg=0.35 wt% (9,11)). Note that Mg is an important constituent of hydrotalcite–like phases ($[Mg_{1-x}Al_x(OH)_2]^{x+}(A^{n-})_{x/n} \cdot yH_2O$), which were also observed in the hydrating cement matrix (9) and, belong to the group of the LDH-phases. Aluminate phases ($3CaO \cdot Al_2O_3$) were found to readily dissolve, thereafter, being the main source of Al in the cement matrix. Note, however, that Mg was not observed to be present in the aluminate phases (11). Therefore, it is speculated that Mg could play an important role in the formation of Ni-Al LDH and, thus, in the Ni accumulation in the reactive zones around alite. Mg may be bound in the neo-formed Ni-Al LDH, but at concentrations well below the detection limit of the techniques used in the present study.

The regions studied by SEM were also investigated by µ-XRF. Fig. 4.3a shows the elemental distribution of Ni, Ca and Fe in a ~1000x1000 µm^2 overview map. The region shows a heterogeneous distribution of these elements. All elements reveal higher and lower concentrated regions. The regions with high Ca or Fe concentration represent the clinker minerals, whereas the lower concentrated regions represent the hydrated cement phases. Ni shows an anti-correlation with Ca and no correlation with Fe, consistent with the SEM/EDS results. The SEM region outlined in Fig. 4.1 represents region 1 on the µ-XRF map (Fig. 4.3a). The area outlined in Fig. 4.2 represents region 2 (Fig. 4.3a). The observation that µ-XRF maps display Ni 'spots-like' instead of 'rims', is a consequence of the penetration depth. The higher penetration depth of the synchrotron X-ray beam (~hundred microns) compared to the penetration depth of an electron beam (a few microns) caused the fluorescence signal to smear out. The Ni speciation was determined on various areas, herein referred to as ROI (region of interest) at both high and low Ni concentrated regions using µ-extended X-ray absorption fine structure (µ-EXAFS) spectroscopy.

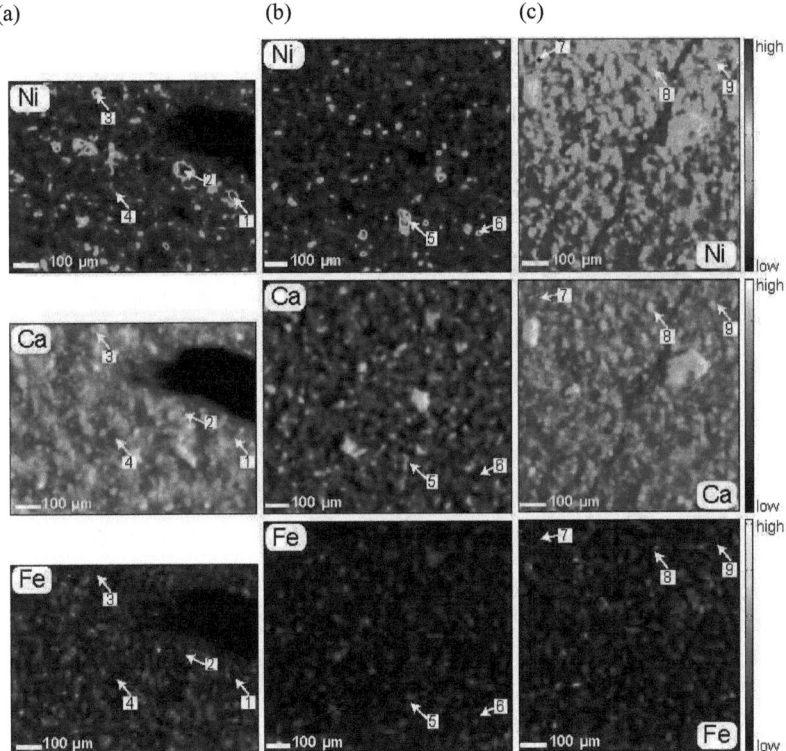

Fig. 4.3. µ-XRF elemental distribution maps of Ni, Ca and Fe for the Ni(NO$_3$)$_2$-doped HCP samples with a water/cement ratio of 0.4, a hydration time of 30 days and a final metal concentration of a) 5000 mg/kg (the black region is silver, used as marker), b) 500 mg/kg, c) 50 mg/kg. Selected regions for µ-EXAFS/µ-XANES measurements are marked 1 to 9. The color bar represents relative concentrations in each sample. Note that region 1 represents the Ni-rich rim in Fig. 4.1, whereas region 2 represents the Ni-rich region shown in Fig. 4.2.

Fig. 4.4 shows selected µ-EXAFS measurements collected at single regions together with the relevant reference spectra. The normalized, background-subtracted and k^3-weighted µ-EXAFS spectra of Ni(NO$_3$)$_2$-doped HCP samples with metal concentration of 50, 500, 5000 mg/kg) and hydrated for 30 days are shown in Fig. 4.4a. Most μ-EXAFS measurements were performed on the HCP sample with the highest metal concentration (Ni_cem_5000). All μ-EXAFS experimental spectra of this sample show similar features. The oscillation at ~4 Å^{-1} is broad and its position is similar to that of the neo-formed LDH phase, which forms upon Ni uptake by pyrophyllite (N-LDH) (Fig. 4.4a). A small feature at ~5 Å^{-1} appears in the experimental spectra of ROI 1 and 3, which is well reproduced in both synthetic and neo-formed Ni-Al LDH spectra. The beat pattern at ~8 Å^{-1} shows a splitting of the oscillation, which is better visible in the Fourier-backtransformed (FT^{-1}) spectra (Fig. 4.4c). Scheinost and Sparks (12) demonstrated that the beat pattern at ~8 Å^{-1} is an indication for the presence of Ni-Al LDH. In fact, this beat pattern is observed in both Ni-Al LDH spectra, whereas the other reference compounds (α-Ni(OH)$_2$, β-Ni(OH)$_2$ and Ni-phyllosilicate) show an elongated upward oscillation ending in a sharp tip at ~8.5 Å^{-1}. Thus, the presence of the beat pattern at ~8 Å^{-1} together with the observed spectral features at ~4 Å^{-1} and ~5 Å^{-1} clearly indicate the presence of a Ni-Al LDH phase. These features appear in all collected spectra in the Ni-doped HCP sample with Ni concentrations of 5000 mg/kg.

The corresponding Fourier Transforms (FT) of the k^3-weighted μ-EXAFS spectra are shown in Fig. 4.4b. The position of the first and second peak of the FT and the shape of the imaginary part reveal strong similarities with the Ni-Al LDH compounds. It should be noted that the amplitude of the second peak in the single experimental spectra is clearly reduced. This observation also applies for the Ni-Al LDH reference compounds, but not for the other references ((5,12); see below). This finding further corroborates the presence of Ni-Al LDH at the Ni-rich regions in the Ni-doped HCP sample.

The structural parameters derived from multi-shell analysis (R+ΔR range=0.8-3.5 Å) are summarized in Table 4.1. Data analysis reveals similar CN and interatomic distances (R) for all investigated regions (Table 4.1, Table S4.2, Supporting Information). The first FT peak corresponds to an octahedral coordination of Ni with ~6 oxygen atoms at 2.03-2.06 Å. The second FT peak reveals strongly reduced CN_{Ni-Ni} (~3) compared to α-Ni(OH)$_2$ (~5) and β-Ni(OH)$_2$ (~6) (5). The CN_{Ni-Ni} of the experimental spectra are comparable to those determined for Ni-Al LDH. The CN_{Ni-Ni} is reduced as Ni is partly substituted by Al in Ni-Al LDH. This causes a significant destructive interference between Ni and Al EXAFS contributions, which results in an amplitude cancellation of the Ni and Al shells (5). Although the CN_{Ni-Ni} of the single ROIs of the Ni-doped HCP sample and Ni-Al LDH agree very well, the overall Ni-Ni distances (3.08-3.11 Å) are significantly longer than those in Ni-Al LDH (3.06-3.07 Å) (5).

This finding suggests that, in addition to Ni-Al LDH, other Ni-containing phases are present. The longer R_{Ni-Ni} is attributed to the presence of β-Ni(OH)$_2$ impurities (R_{Ni-Ni}=3.13 Å) (5).

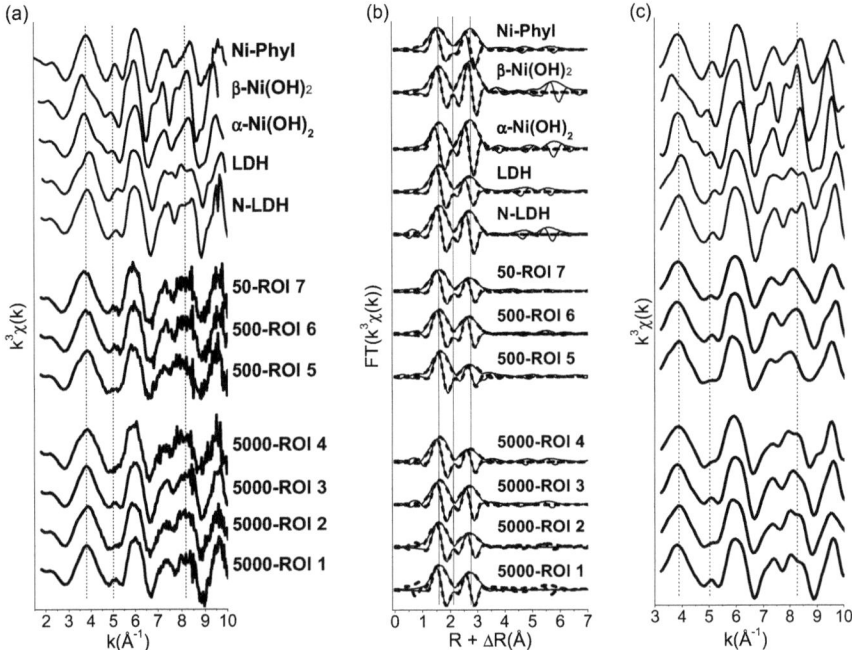

Fig. 4.4. Ni K-edge spectra for Ni reference compounds and selected μ-EXAFS experimental spectra of the Ni(NO$_3$)$_2$-doped HCP samples with a water/cement ratio of 0.4, a hydration time of 30 days and three different total metal concentrations (50, 500, 5000 mg/kg), a) k^3-weighted, normalized, background-subtracted spectra, b) Experimental (solid line) and calculated (dashed line) Fourier Transforms (modulus and imaginary parts) obtained from the μ-EXAFS spectra presented in a), c) k^3-weighted EXAFS data for the Fourier-backtransform spectra obtained from Fig. 4.4b (R+ΔR range=0.8-7 Å). Dotted lines indicate spectral features explained in detail in the text. Notations: ROI = region of interest, N-LDH = neo-formed Ni-Al LDH, LDH = synthetic Ni-Al LDH (Ni:Al, 2:1 (8)), Ni-Phyl = Ni-phyllosilicate.

Table 4.1. Structural information obtained from selected μ-EXAFS Ni K-edge data analysis (region of interest, ROI, are indicated in Fig. 4.3).

Cement Samples	Ni-O			Ni-Ni			ΔE_0	%Res
	CN	R (Å)	σ^2 (Å)	CN	R (Å)	σ^2 (Å)		
Ni_cem_5000 (bulk-EXAFS)[a]	7.3	2.05	0.007	3.0	3.11	0.005[b]	1.3	7.9
ROI 1	5.1	2.05	0.005	3.0	3.09	0.005[b]	0.7	9.3
ROI 2	5.7	2.04	0.006	2.5	3.09	0.005[b]	-1.9	6.1
ROI 3	4.7	2.05	0.003	3.2	3.09	0.005[b]	0.2	9.2
ROI 4	6.6	2.06	0.007	2.3	3.10	0.005[b]	-0.4	9.9
Ni_cem_500 (bulk-EXAFS)[a]	5.3	2.05	0.005[b]	2.3	3.12	0.005[b]	0.0	9.0
ROI 5	5.3	2.08	0.005	2.0	3.12	0.005[b]	3.2	11.6
ROI 6	6.2	2.05	0.007	2.9	3.07	0.005[b]	-1.2	5.8
Ni_cem_50 (bulk-EXAFS)[a]	6.3	2.04	0.005[b]	1.8	3.17	0.005[b]	-0.3	10.8
ROI 7	5.7	2.05	0.007	2.3	3.09	0.005[b]	0.6	5.2

[a] Vespa et al., 2006, [b] fix parameters during fitting procedures
R, CN, σ^2, ΔE_0 stand for interatomic distances, coordination numbers, Debye-Waller factors and inner potential corrections.
Estimated error: $R_{(Ni-O)}$ ±0.02 Å, $CN_{(Ni-O)}$ ±20%, $R_{(Ni-Ni)}$ ±0.02 Å, $CN_{(Ni-Ni)}$ ±20%
%Res: deviation between experimental data and fit given by the relative residual in percent.
N= number of data points, Y_{exp} and Y_{theo}: experimental and theoretical data points, respectively.

$$\% Res = \frac{\sum_{i=1}^{N}|y_{exp}(i) - y_{theo}(i)|}{\sum_{i=1}^{N} y_{exp}(i)} *100$$

4.3.2 Influence of varying Ni concentrations

The macro-spectroscopic study of Vespa et al. (5) demonstrated that, at a high Ni loading of 5000 mg/kg, a mixture of different Ni phases forms. Ni-Al LDH was found to be the predominant Ni species, which precipitates together with a small amount of α-Ni(OH)$_2$ and β-Ni(OH)$_2$. The above presented data from the same HCP sample shows that, on the micro-scale, mainly Ni-Al LDH forms together with small amounts of β-Ni(OH)$_2$. The findings from the micro-scale investigations are in good agreement with the results from the macro-spectroscopic study (5), indicating that the same Ni species form and, in particular Ni-Al LDH is the predominant Ni phase in the cement matrix on both the macro- and micro-scale.

To address the question whether the findings from the Ni-doped HCP sample with 5000 mg/kg Ni loading can be extrapolated to samples with lower Ni concentrations, samples with 50 and 500 mg/kg Ni loadings have been investigated. The BSE images of the Ni-doped HCP samples with Ni concentrations of 50 and 500 mg/kg (Figure S4.1, Supporting Information) indicate that the microstructure of HCP is comparable to that of the sample with 5000 mg/kg Ni loading (Fig. 4.1 and Fig. 4.2). In fact, the Ni phases always form rims around the inner-C-S-H phase independent of the loading. The EDS analysis further reveals that Ni is not present in clinker minerals and hydrated cement phases (Table S4.1, analyses r and w, x indicated in Figure S4.1). In addition, the Ni concentrations on single Ni-rich spots on these rims vary only slightly between ~3-~5 wt% (Table S4.1, analyses s-v indicated in Figure S4.1) for the sample with 500 mg/kg and ~1-~3wt% (Table S4.1, analyses y-z indicated in Figure S4.1) for the sample with 50 mg/kg Ni loading. This indicates that, although the total Ni concentration in these samples is 10 and 100 times lower than in the Ni-doped HCP sample with 5000 mg/kg, the precipitation of a Ni phase still takes place. The BSE/EDS investigations were complemented with μ-XRF measurements. The μ-XRF distribution maps for the samples with 50 and 500 mg/kg Ni loadings indicate a heterogeneous distribution of Ni, Ca and Fe as previously observed for the sample with the higher Ni loading (5000 mg/kg) (Fig. 4.3). The Ni speciation was determined at selected Ni-rich regions in those samples using μ-EXAFS. Fig. 4.4 shows a comparison of the normalized, background-subtracted and k^3-weighted μ-EXAFS spectra (a), the FT (b), and the FT^{-1} (c) for the various ROIs. The μ-EXAFS spectra of the sample with 50 and 500 mg/kg are similar to each other and to the previously discussed μ-EXAFS spectra of the sample with 5000 mg/kg. Data analysis reveals a first coordination shell with ~6 oxygen at R_{Ni-O} of 2.04-2.06 Å for all samples (Table 4.1). For the second coordination shell CN is ~3, which is in good agreement with CN_{Ni-Ni} in Ni-Al LDH. Bond distances for the second shell (R_{Ni-Ni}) are determined to be 3.07-3.12 Å, which corresponds to distances previously determined for the sample with 5000 mg/kg Ni loading. This finding suggests that the composition of the Ni phases formed in Ni-rich regions is similar, regardless of the metal loading of HCP. Furthermore, the μ-EXAFS results of the samples with 500 and 5000 mg/kg Ni loadings agree very well with the findings from the macro-spectroscopic studies (5), indicating that the Ni-rich phases observed on the micro-scale are relevant to the whole matrix.

In the macro-spectroscopic study, Vespa et al. (5) indicated that the Ni speciation was different for the sample with initial Ni concentration of 50 mg/kg. The study showed that the structural model used to analyze the data collected at increased Ni concentrations (>500 mg/kg) is not valid at the lowest Ni concentration (50 mg/kg). In fact, the fit approach resulted in Ni-Ni distances longer (3.17 Å) than any known Ni compound (<3.13 Å). The μ-EXAFS measurements aimed at investigating whether the Ni speciation deduced from the bulk-EXAFS on the 50 mg/kg Ni-doped HCP sample (5) can be discerned on the micro-scale. Fig. 4.3c reveals a heterogeneous Ni distribution with Ni-rich and Ni-poor regions for the 50 mg/kg Ni-doped HCP sample. The μ-EXAFS results of the Ni-rich ROI 7 (Fig. 4.4; Table 4.1) indicate the formation of Ni-Al LDH, as discussed above. Thus, the results obtained for

Ni-rich regions from the HCP sample at 50 mg/kg Ni loading are comparable with the results determined for the HCP samples at higher Ni concentrations. These findings suggest that Ni-Al LDH forms even in the HCP sample at 50 mg/kg Ni loading. Note, however, that the observation of Ni-Al LDH in the HCP sample at this low Ni concentration was not supported by the bulk-EXAFS measurements of the same sample (5). In latter studies, Ni-Al LDH was not the only species present in the cement matrix. To further discern the nature of the predominant Ni species in the 50 mg/kg Ni-doped HCP sample, Ni-poor regions were investigated. Due to the low metal concentration and the long acquisition time needed under these conditions, it was only possible to collect µ-X-ray absorption near edge structure (µ-XANES) and µ-EXAFS data up to 7 Å^{-1} at Ni-poor regions of the cement matrix. Fig. 4.5a shows normalized µ-XANES spectra of ROIs 8 and 9 together with those determined for the Ni-rich ROI 7 (indicated in Fig. 4.3c) and the corresponding bulk-XANES. The figure clearly reveals that the spectra collected at the Ni-poor ROIs 8 and 9 and the bulk-XANES data are comparable. However, the spectrum collected at the Ni-rich ROI 7 is different, showing two distinct features. The first feature is observed at 8350 eV. The second feature appears at 8390 eV, where the spectra of ROIs 8 and 9 show a splitting of the oscillation. This feature is also seen in the bulk-XANES spectrum and, clearly visible in the µ-EXAFS spectra at ~4 Å^{-1} (Fig. 4.5b). These findings suggest that, at low Ni concentration at least two different Ni species are present. In strongly Ni-enriched regions (e.g., ROI 7) Ni-Al LDH forms, probably mixed with β-Ni(OH)$_2$. At selected regions with significantly lower Ni concentrations (e.g., ROI 8 and 9), however, an additional Ni species is observed. This unknown Ni species is neither a Ni-Al LDH nor any other known Ni precipitate. Although the structural parameters of the species could not be identified in the present study, the species dominates in the cement matrix, as indicated from the previous bulk-EXAFS investigations (5).

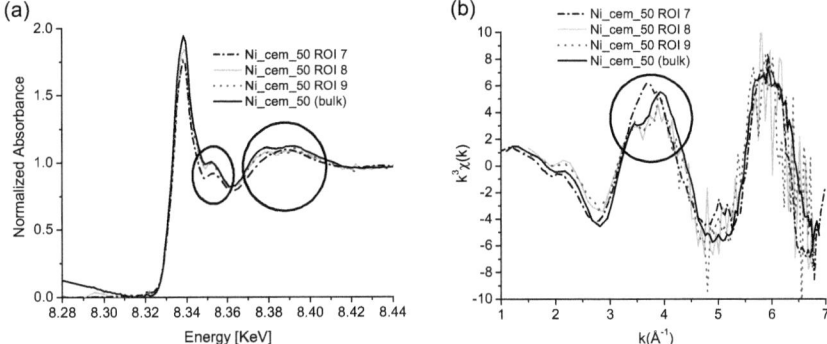

Fig. 4.5. Ni K-edge μ-XANES experimental spectra of ROIs (regions of interest) 7, 8 and 9 shown in Fig. 4.3c of the Ni(NO₃)₂-doped HCP sample with a water/cement ratio of 0.4, a hydration time of 30 days and a total metal concentration of 50 mg/kg, a) normalized absorbance, b) k^3-weighted, normalized, background-subtracted μ-EXAFS spectra. The powder experimental spectrum of the same sample (Ni_cem_50; (5)) is shown for comparison. Circles indicate spectral features explained in detail in the text.

Supporting Information Available

Table S4.1 shows various EDS analyses and Table S4.2 shows additional μ-EXAFS results of the Ni-doped HCP samples.

Figure S4.1 shows BSE images and the regions of the EDS analysis of two Ni-doped HCP samples with low Ni-loadings (500 and 50 mg/kg).

4.4 References

(1) Schmidt, M.; Beckefeld, P.; Götz, R.; Kamsties, S.; Kretz, C.; Molitor, N.; Neck, U.; Vogel, P. *Reststoff-und Abfallverfestigung. Immobilisierung von Schadstoffen-Recycling-Verbesserung der Deponiefähigkeit*; Expert Verlag: Renningen-Malmheim, **1995**.

(2) Chapman, N.; McCombie, C. *Principles and standards for the disposal of long-lived radioactive wastes*; 1 ed.; Elsevier Science, Ltd.: Oxford, **2003**.

(3) Glasser, F. P. Chemistry of cement-solidified waste forms. . In: *Chemistry and microstructure of solidified waste forms*; Spence, R. D., Ed.; Lewis Publishers: Boca Raton, **1993**.

(4) Scheidegger, A. M.; Wieland, E.; Dähn, R.; Spieler, P. Spectroscopic evidence for the formation of layered Ni-Al double hydroxides in cement. *Environmental Science and Technology* **2000**, *34*, 4545-4548.

(5) Vespa, M.; Dähn, R.; Grolimund, D.; Wieland, E.; Scheidegger, A. M. Spectroscopic investigation of Ni speciation in hardened cement paste. *Environmental Science and Technology* **2006**, *40*, 2275-2282.

(6) Marcus, M.; MacDowell, A. A.; Celestre, R.; Manceau, A.; Miller, T.; Padmore, H. A.; Sublett, R. E. Beamline 10.3.2 at ALS: A hard- X-ray microprobe for environmental and material sciences. *Journal of Synchrotron Radiation* **2004**, *11*, 239-247.

(7) Ressler, T. WinXAS: A program for X-ray absorption spectroscopy data analysis under MS-Windows. *Journal of Synchrotron Radiation* **1998**, *5* (2), 118-122.

(8) Johnson, C. A.; Glasser, F. P. Hydrotalcite-like minerals ($M_2Al(OH)_6(CO_3)_{0.5} \cdot XH_2O$, where M= Mg, Zn, Co, Ni) in the environment: synthesis, characterization and thermodynamic stability. *Clays and Clay Minerals* **2003**, *51*, 1-8.

(9) Lothenbach, B.; Wieland, E. A thermodynamic approach to the hydration of sulphate-resisting portland cement. *Waste Management* **2006**, *26*, 706-719.

(10) Peltier, E.; Allada, R.; Navrotsky, A.; Sparks, D. L. Nickel solubility and precipitation in soils: A thermodynamic study. *Clays and Clay Minerals* **2006**, *54* (2), 153-164.

(11) Taylor, H. F. W. *Cement Chemistry*; 2nd ed.; Thomas Telford: London, **1997**.

(12) Scheinost, A. C.; Sparks, D. L. Formation of layered single- and double-metal hydroxide precipitates at the mineral/water interface: A multiple-scattering XAFS analysis. *Journal of Colloid and Interface Science* **2000**, *223*, 1-12.

SUPPORTING INFORMATION FOR CHAPTER 4

Micro-scale investigations of Ni uptake by cement using a combination of scanning electron microscopy and synchrotron-based techniques

S4.1 Details of μ-XAS data reduction

All data were processed following the same procedure. The energy was converted to photoelectron wave vector units ($Å^{-1}$) by assigning the origin E_0 to the first inflection point of the absorption edge. Radial Structure Functions (RSF) were obtained by Fourier transforming the k^3-weighted $\chi(k)$ functions between 3.2 and 10.9 $Å^1$ with a Bessel window function with a smoothing parameter of 4. Multishell fits were performed in real space across the range of the first two shells (R+ΔR range=0.8-3.5 Å). The first coordination shell was fitted with Ni-O backscattering pairs. The second coordination shell was fitted solely using Ni-Ni pairs, because the discrimination of Ni-Ni and Ni-Al backscattering pairs in Ni-Al LDH is problematic (1). To be able to compare the coordination numbers (CN) of the Ni-Ni backscattering pairs (CN_{Ni-Ni}) of all samples and references, the Debye-Waller factor was set to 0.005 $Å^2$ (2). Theoretical scattering paths for the fit were calculated using FEFF 8.20 (3) and the structure of β-Ni(OH)$_2$ as a reference. The amplitude reduction factor (S_0^2) was determined to be 0.85 from the experimental β-Ni(OH)$_2$ XAS spectrum (1,2).

Table S4.1. Semi-quantitative EDS analyses of Ni-rich regions and cement phases in the Ni-doped HCP-samples. All data are given in wt% (spots are indicated in Fig. 1b, Fig. 2b and Figure S4.1).

Ni_cem_5000	Ni	Al	Ca	Si	Fe	S	Sr	Mg	K	Na	Cr	Cl	O	Total
a	0.00	0.37	34.92	11.37	0.68	1.20	0.00	0.18	0.06	0.07	0.00	0.01	51.14	100.00
b	0.13	0.03	53.86	1.10	0.03	0.19	0.00	0.06	0.02	0.09	0.04	0.00	44.45	100.00
c	28.33	3.33	10.21	1.85	0.12	1.17	0.58	0.00	0.03	0.00	0.05	0.04	54.31	100.02
d	27.42	3.44	10.54	2.09	0.17	1.01	0.42	0.17	0.01	0.10	0.06	0.01	54.58	100.02
e	22.50	3.54	15.71	4.80	0.41	0.90	1.28	0.38	0.02	0.08	0.00	0.05	50.33	100.00
f	9.22	1.12	34.42	6.66	0.53	0.56	0.00	0.37	0.07	0.13	0.00	0.00	46.92	100.00
g	2.46	0.88	33.54	9.37	0.78	1.23	0.00	0.34	0.04	0.00	0.10	0.13	51.14	100.01
h	5.51	1.48	28.57	9.95	0.43	1.11	0.00	0.72	0.04	0.00	0.00	0.02	52.17	100.00
i	0.24	0.91	24.03	8.94	0.47	0.78	0.00	0.24	0.05	0.00	0.02	0.00	64.33	100.01
j	0.00	0.37	21.25	8.91	0.27	0.63	0.00	0.47	0.00	0.12	0.04	0.03	67.91	100.00
k	0.01	0.23	29.64	9.06	0.06	0.11	0.00	0.27	0.04	0.17	0.00	0.02	60.40	100.01
l	10.85	1.92	25.63	8.43	0.48	0.79	1.96	0.43	0.01	0.08	0.00	0.03	49.41	100.02
m	7.51	1.44	26.10	8.21	0.48	0.58	1.88	0.63	0.00	0.28	0.05	0.00	52.85	100.01
n	4.23	1.54	33.89	6.96	1.16	0.77	1.70	0.51	0.00	0.00	0.01	0.05	49.16	99.98
o	5.97	1.43	38.21	8.08	1.07	0.78	2.02	0.38	0.06	0.00	0.08	0.08	41.84	100.01
p	13.16	2.46	24.38	8.49	0.32	0.70	2.02	0.42	0.07	0.00	0.02	0.02	47.96	100.02
q	12.27	2.29	25.81	8.27	0.47	0.96	2.01	0.30	0.02	0.00	0.07	0.07	47.45	99.99
Ni_cem_500														
r	0.24	0.49	36.18	10.67	0.74	0.95	2.62	0.79	0.00	0.05	0.00	0.02	47.25	100.00
s	2.51	0.65	34.25	9.67	0.42	1.28	2.48	0.52	0.17	0.05	0.06	0.13	47.80	99.99
t	4.00	0.77	33.52	9.70	0.37	0.92	2.29	0.50	0.17	0.10	0.07	0.04	47.56	100.01
u	4.45	0.89	31.35	9.22	0.49	1.06	2.18	0.70	0.15	0.00	0.00	0.00	49.52	100.01
v	2.52	0.65	32.16	9.29	0.34	0.93	2.19	0.55	0.15	0.55	0.00	0.07	50.60	100.00
Ni_cem_50														
w	0.21	0.13	55.51	2.06	0.14	0.59	0.20	0.03	0.05	0.14	0.05	0.03	40.85	99.99
x	0.17	0.86	37.46	10.72	1.08	1.15	0.00	0.58	0.06	0.12	0.01	0.00	47.81	100.02
y	3.08	1.26	34.17	10.12	1.18	1.49	0.00	0.65	0.08	0.00	0.07	0.06	47.84	100.00
z	1.12	1.02	38.33	10.41	1.24	1.54	0.00	0.32	0.09	0.09	0.00	0.09	45.74	99.99

errors: ~1 wt%

Table S4.2. Structural information obtained from further selected μ-EXAFS Ni K-edge data analysis showed together with reference compounds (4,5) and bulk-EXAFS data (2).

Samples	Ni-O			Ni-Ni			Ni-Si			ΔE_0	%Res
	CN	R (Å)	σ^2 (Å)	CN	R (Å)	σ^2 (Å)	CN	R (Å)	σ^2 (Å)		
References											
Ni-Phyllosilicate [a]	5.1	2.04	0.006	3.5	3.07	0.008[d]	3.7	3.26[SI]	0.008[d]	0.3	3.0
β-Ni(OH)$_2$	5.6	2.06	0.005	5.6	3.13	0.005[e]				-0.6	3.0
α-Ni(OH)$_2$	5.2	2.03	0.005	4.9	3.09	0.005[e]				3.0	4.4
Ni-Al LDH (LDH)	6.0	2.05	0.006	2.5	3.06	0.005[e]				1.1	4.5
Neo-formed Ni-Al LDH [b] (N-LDH)	5.7	2.04	0.004	3.9	3.07	0.005[e]				0.3	3.8
Cement samples											
Ni_cem_5000 (bulk-EXAFS)[c]	*7.3*	*2.05*	*0.007*	*3.0*	*3.11*	*0.005[e]*				*1.3*	*7.9*
ROI S1	5.9	2.06	0.006	3.5	3.09	0.005[e]				1.5	4.8
ROI S2	4.6	2.06	0.002	3.0	3.11	0.005[e]				2.1	7.3
ROI S3	6.6	2.04	0.007	3.1	3.08	0.005[e]				-0.7	3.1
ROI S4	5.3	2.03	0.003	2.6	3.08	0.005[e]				-2.8	11.2

[a] Dähn et al. 2002, [b] Scheidegger et al., 1997, [c] Vespa et al. (2006), [d] correlated parameters and [e] fix parameters during fitting procedures
R, CN, σ^2, ΔE_0 stand for interatomic distances, coordination numbers, Debye-Waller factors and inner potential corrections.
Estimated error: R$_{(Ni-O)}$ ±0.02 Å, CN$_{(Ni-O)}$ ±20%, R$_{(Ni-Ni)}$ ±0.02 Å, CN$_{(Ni-Ni)}$ ±20%
%Res: deviation between experimental data and fit given by the relative residual in percent.
N= number of data points, Y$_{exp}$ and Y$_{theo}$: experimental and theoretical data points, respectively.

$$\% Res = \frac{\sum_{i=1}^{N}|y_{exp}(i) - y_{theo}(i)|}{\sum_{i=1}^{N} y_{exp}(i)} *100$$

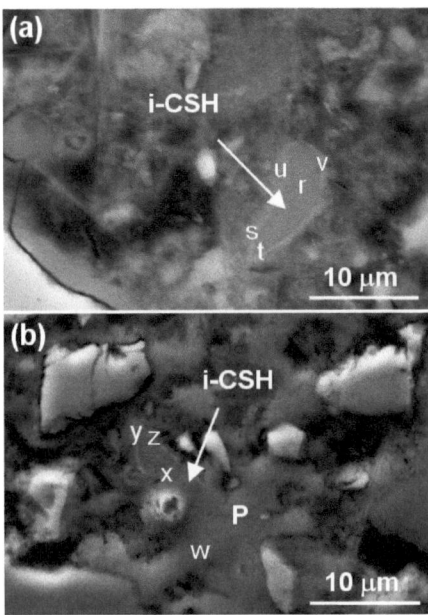

Figure S4.1. BSE images of the Ni(NO$_3$)$_2$-doped HCP sample with a water/cement ratio of 0.4 hydrated for 30 days and total metal concentrations of (a) 500 mg/kg and (b) 50 mg/kg. The lower case letters s to z indicate the regions of the SEM/EDS spot analyses shown in Table S4.1.

S4. References

(1) Scheidegger, A. M.; Wieland, E.; Dähn, R.; Spieler, P. Spectroscopic evidence for the formation of layered Ni-Al double hydroxides in cement. *Environmental Science and Technology* **2000**, *34*, 4545-4548.

(2) Vespa, M.; Dähn, R.; Grolimund, D.; Wieland, E.; Scheidegger, A. M. Spectroscopic investigation of Ni speciation in hardened cement paste. *Environmental Science and Technology* **2006**, *40*, 2275-2282.

(3) Rehr, J. J.; Albers, R. C. Theoretical approaches to X-ray absorption fine structure. *Reviews of Modern Physics* **2000**, *72* (3), 621-653.

(4) Dähn, R.; Scheidegger, A. M.; Manceau, A.; Schlegel, M. L.; Baeyens, B.; Bradbury, M. H.; Morales, M. Neoformation of Ni phyllosilicate upon Ni uptake on montmorillonite: A kinetics study by powder and polarized extended X-ray absorption fine structure spectroscopy. *Geochimica et Cosmochimica Acta* **2002**, *66* (13), 2335-2347.

(5) Scheidegger, A. M.; Lamble, G. M.; Sparks, D. L. The kinetics of nickel sorption on phyrophyllite as monitored by X-ray absorption fine structure (XAFS) spectroscopy. *Journal de Physique IV France* **1997**, *7* (C2), 773-775.

CHAPTER 5

THE INFLUENCE OF VARYING HYDRATION TIME ON THE NI UPTAKE BY CEMENT

Abstract

In this study, Ni uptake by hardened cement paste has been investigated with the aim of improving our understanding of the immobilization process of Ni(II) in cement and the influence of the hydration time on the Ni speciation on the microscopic scale. Information on the Ni distribution and speciation of the Ni phases formed in the cement system has been gained by employing synchrotron-based µ-X-ray fluorescence (µ-XRF) and µ-X-ray absorption spectroscopy (µ-XAS). The Ni-doped cement samples were prepared at a water/cement ratio of 0.4 and with a metal loading of 5000 mg/kg using a sulphate-resisting Portland cement. The samples were hydrated for six hours and one year to account for the chemical environment in a fresh and aged cement paste, respectively. The µ-XAS measurements reveal that a mixture of Ni phases form at single regions of interests, independent of the hydration time. Data analysis further indicates that Ni(II) is predominantly immobilized in a layered double hydroxide (LDH, Ni-Al LDH) and only to a minor extent precipitates as Ni-hydroxides. A comparison of the results from this micro-spectroscopic investigation with those from an earlier macro-spectroscopic study, indicate that the same Ni phases form both on the macro- and micro-scale.

5.1 Introduction

Cement-based materials have been used worldwide for conditioning radioactive wastes in order to prevent or lower the mobility of the contaminants present in the waste matrices. To ensure the long-term safe disposal of cement-conditioned radioactive waste a deep geological disposal is foreseen for some categories of waste forms (1). In the radioactive waste, Ni radioisotopes are mainly associated with irradiated metallic components from nuclear power plants. A molecular-level understanding of the processes governing the immobilization of radionuclides in hydrating cement is essential for detailed predictions of their long-term release from the cementations near field of a repository for radioactive waste. Hardened cement paste (HCP) is a very heterogeneous material with discrete particles in the nano- to micrometer size range. The material consists of mainly calcium (aluminum) silicate hydrates (C-S-H), portlandite (calcium hydroxide) and calcium aluminates. In addition, small amounts of non-hydrated clinker minerals (alite, belite, ferrite aluminate) may be present. The immobilization potential of HCP originates from its selective binding properties for metal cations and anions (e.g. (2)). Thus, it appears that immobilization processes in cement systems are highly specific with respect to the mineral components and mechanisms involved.

In the past, several X-ray absorption spectroscopy (XAS) studies have been reported on the Ni uptake by cement. For example, Scheidegger et al. (3,4) investigated the sorption of Ni onto hydrated Portland cement. The study showed that a mixture of $Ni(OH)_2$ and Ni-Al layered double hydroxide (Ni-Al LDH, $[M^{II}_{1-x}M^{III}_x(OH)_2]^{x+}(A^{n-})_{x/n} \cdot yH_2O$) phases was formed. In a recent study, Vespa et al. (5) investigated the Ni speciation during cement hydration by combining XAS measurements on powdered Ni-doped cement material. The authors varied important experimental parameters such as the hydration time, the anions added to the system, the water/cement (w/c) ratio and the metal concentrations. The XAS study revealed the predominant formation of Ni-Al LDH in the Ni-doped cement system and that Ni-hydroxide phases (α-$Ni(OH)_2$ and β-$Ni(OH)_2$) were present only to a minor extent. The study further showed that the amount of Ni-Al LDH increased as a function of the hydration time, whereas α-$Ni(OH)_2$ decreased and β-$Ni(OH)_2$ remained constant. The only exception was found in the sample with the lowest Ni-loading (50 mg/kg), in which the precipitation of Ni-Al LDH and Ni-hydroxides phases were not the predominant Ni species. In a further study, Vespa et al. (6) investigated the influence of the inherent heterogeneity of the cement matrix on the Ni speciation on the micro-scale. The authors focused on the influence of the different Ni loadings on the Ni speciation in the Ni-doped cement sample hydrated for 30 days. The study demonstrated that the same Ni speciation observed on the macro-scale is also found on the micro-scale, independent of the Ni-loadings used. This finding indicates that the same Ni species are relevant both on the micro-scale and for the whole cement matrix.

To further address the question concerning the influence of the hydration time on the Ni speciation micro-level synchrotron-based μ-X-ray fluorescence (μ-XRF) and μ-extended X-ray absorption fine structure (μ-EXAFS) were carried out on Ni-doped cement samples, which were hydrated for six hours and one year. This combined approach enables to obtain spatially-resolved information on the Ni distribution and on the chemical speciation of Ni at selected regions. This information will allow developing a molecular-level understanding of the Ni binding in cement.

5.2 Materials and methods
5.2.1. Sample preparation

The cement samples were prepared from a commercial sulphate-resisting Portland cement (CEM I 52.5 N HTS, Lafarge, France) used to condition radioactive waste in Switzerland. Ni-doped HCP was prepared by mixing a $Ni(NO_3)_2$ solution with unhydrated cement. The metal salt was dissolved in deionized water to obtain stock solutions with concentrations of 0.3 mol/L (pH=4.5). The solutions were mixed with the unhydrated cement at a w/c ratio of 0.4, using a standard procedure (7). The final metal concentration of the pastes was 5000 mg/kg dry HCP. The cement pastes were stored into plexiglass moulds, which were closed with polyethylene lids, and hydrated for six hours and one year. For the short hydration of six hours, the slurry was filtered (0.2 μm pore size) to separate the solid from the free water. The solid material was washed with acetone for 15 minutes to stop the hydration process (8), filtered and dried in a glovebox under controlled N_2 atmosphere (CO_2, O_2<2 ppm, T=20±3 °C). The sample hydrated for one year was stored in a closed container at room temperature at 100% relative humidity. After hydration was completed, the cylinder was cut into several slices of ~1 cm thickness and dried in the glovebox in a dry N_2 atmosphere. Selected slices from the one year and six hours hydrated samples were impregnate and polished for the preparation of thin sections, which were employed for μ-XRF and μ-EXAFS measurements.

5.2.2. µ-XRF and µ-EXAFS data collection and reduction

µ-XRF and µ-EXAFS data were collected on beamline 10.3.2 at the Advanced Light Source (ALS), Berkeley, USA (9). The beamline is equipped with a Si(111) crystal monochromator. The monochromator angle was calibrated by assigning the energy of 8333 eV to the first inflection point of the K-edge absorption spectrum of Ni metal foil. The µ-XRF and µ-EXAFS measurements were collected at room temperature in fluorescence mode with a 7 element Ge-solid state detector with a beam size of ~5x5 µm^2. The µ-XRF maps were obtained by scanning the sample under the monochromatic beam at an energy of 10000 eV with a pixel size of 5x5 µm^2.

The µ-XRF maps were processed using the Labview software package at the beamline 10.3.2 (9) and MATLAB. µ-EXAFS data at the Ni K-edge were dead-time corrected using the Labview software package at the beamline 10.3.2. Further data reduction was performed using the WinXAS 3.1 software package (10). All spectra were normalized by fitting a first-degree polynomial to the pre-edge and a third-degree polynomial to the post-edge regions. The energy was converted to photoelectron wave vector units (Å^{-1}) by assigning the origin E_0 to the first inflection point of the absorption edge. Radial Structure Functions (RSF) were obtained by Fourier transforming the k^3-weighted χ(k) functions between 3.2 and 10.9 Å^{-1} with a Bessel window function with a smoothing parameter of 4. Multishell fits were performed in real space across the range of the first two shells (R +ΔR range=0.8-3.5 Å). Theoretical scattering paths for the fit were calculated using FEFF 8.20 (11) and the structure of β-Ni(OH)$_2$ as a reference. The amplitude reduction factor (S$_0^2$) was determined to be 0.85 from the experimental β-Ni(OH)$_2$ EXAFS spectrum (3). Errors on the structural parameters were estimated from the analysis of two reference compounds (β-Ni(OH)$_2$, α-Ni(OH)$_2$; see Table 5.1). Several reference spectra (β-Ni(OH)$_2$, α-Ni(OH)$_2$, synthetic Ni-Al LDH (Ni:Al, 2:1; Ni$_2$Al(OH)$_6$(CO$_3$)$_{1/2}$ (12), Ni-phyllosilicate (13), neo-formed Ni-Al LDH formed from Ni-doped pyrophyllite (14)), collected at the Swiss Norwegian Beamline (SNBL) at the ESRF, were used to identify the Ni species in the cement matrix.

5.3 Results and discussion

Fig. 5.1 shows the elemental distribution of Ni and Ca for the 6-hours and 1-year hydrated Ni-doped HCP samples in a ~1000x1000 µm^2 overview map. The regions show a heterogeneous distribution of Ni and Ca. Both elements reveal higher (red for Ni, white for Ca) and lower (blue for Ni, green for Ca) concentrated regions. The regions with high Ca concentration (shown in white) represent the clinker phases, i.e. non-hydrated cement minerals, whereas the lower concentrated regions (shown in green) represent the hydrated cement phases (6). In the HCP sample (Fig. 5.1a) hydrated for six hours the amount of non-hydrated clinker phases (shown in white) is higher than in the HCP sample hydrated for one year (Fig. 5.1b), in which the hydrated cement phases clearly predominate (shown in green). Note that the Ni distribution of these two samples exhibits some differences. It appears that, although both samples have rich (red to yellow areas) and poor (blue areas) Ni regions, the HCP sample hydrated for six hours (Fig. 5.1a) shows a more uniform distribution of the Ni poor regions compared to the HCP sample hydrated for one year. Moreover, the 1-year-hydrated sample (Fig. 5.1b) shows regions in which Ni is depleted (black areas). It also appears that, with increasing hydration time, Ni-rich rims form around Ca-rich particles (clinker minerals) as observed for the µ-EXAFS spot 5 (Fig. 5.1b). By contrast, more homogeneous Ni regions are observed in the HCP sample hydrated for six hours (Fig. 5.1a). Overall, the distribution of Ni reveals a significant anti-correlation with Ca, which is clearly observed in the HCP sample hydrated for one year. Thus, the µ-XRF results, suggest that with increasing hydration time Ni tends to concentrate in specific regions around Ca-rich particles, in particular around the alite clinker mineral as observed from Vespa et al. (6).

The Ni speciation in both samples was determined on various spots at high (red) and low (blue) Ni concentrated regions using µ-EXAFS. Fig. 5.2 shows selected µ-EXAFS measurements collected at single spots together with the relevant reference spectra. The normalized, background-subtracted and k^3-weighted µ-EXAFS spectra of the Ni-doped HCP samples hydrated for six hours and one year are shown (Fig. 5.1a). All µ-EXAFS spectra of these samples show similar features. For all spots the oscillation at ~4 Å^{-1} reveals a similar position to that of the neo-formed LDH phase, which forms upon Ni uptake by pyrophyllite (N-LDH) (14). A small feature at ~5 Å^{-1} appears in all experimental spectra, which is well reproduced in both synthetic and neo-formed Ni-Al LDH spectra. The beat pattern at ~8 Å^{-1} shows a splitting of the oscillation. Scheinost and Sparks (15) demonstrated that the beat pattern at ~8 Å^{-1} is an indication for the presence of Ni-Al LDH. In fact, this beat pattern is observed in both Ni-Al LDH spectra, whereas the other reference compounds (α-Ni(OH)$_2$, β-Ni(OH)$_2$ and Ni-phyllosilicate) show an elongated upward oscillation ending in a sharp tip at ~8.5 Å^{-1}. Thus, the presence of the beat pattern at ~8 Å^{-1}, together with the observed spectral features at ~5 Å^{-1} and ~4 Å^{-1}, clearly indicate the presence of a Ni-Al LDH phase. These features appear in all experimental spectra collected at single spots in both Ni-doped HCP samples, indicating that Ni-Al LDH predominantly forms on the micro-scale.

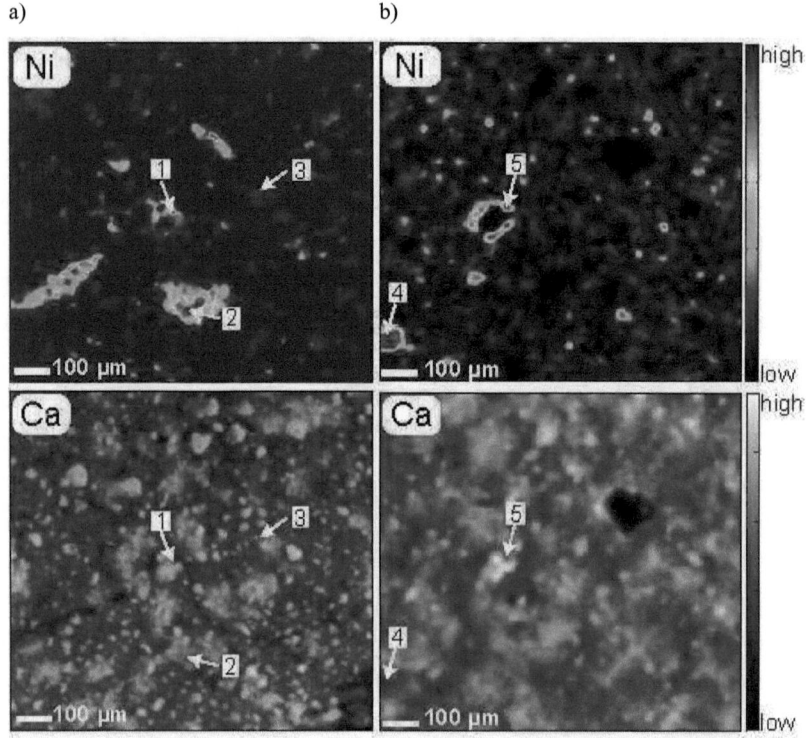

Fig. 5.1 μ-XRF elemental distribution maps of Ni and Ca from the $Ni(NO_3)_2$-doped HCP samples with a water/cement 0.4 and a final metal concentration of 5000 mg/kg with hydration times of a) six hours, and b) one year. Selected spots for μ-EXAFS measurements are marked with numbers from 1 to 5. The color bar represents relative concentrations in each sample.

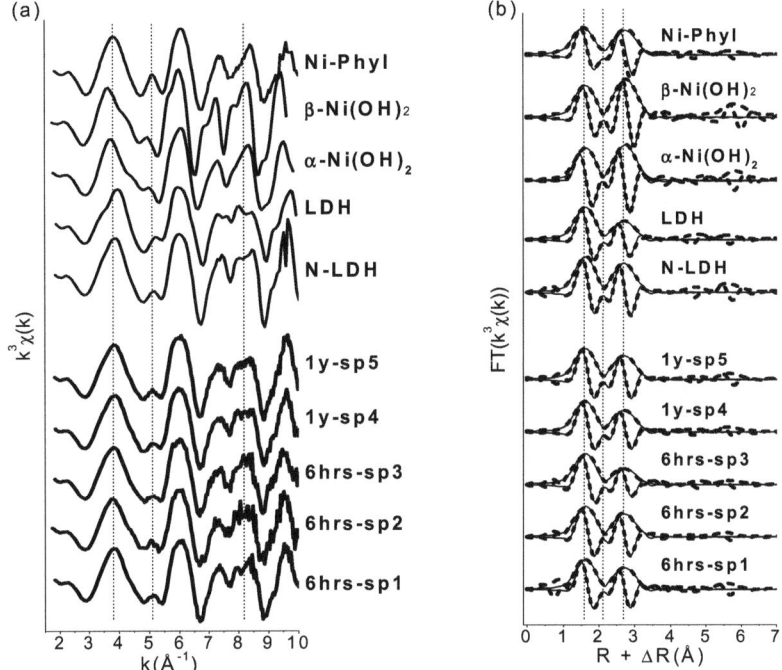

Fig. 5.2. Ni K-edge spectra for Ni reference compounds and selected μ-EXAFS experimental spectra of the Ni(NO₃)₂-doped HCP samples with a water/cement 0.4 and a final metal concentration of 5000 mg/kg hydrated for six hours and one year, a) k^3-weighted, normalized, background-subtracted spectra, b) Experimental (solid line) and calculated (dashed line) Fourier Transforms (modulus and imaginary parts) obtained from the μ-EXAFS spectra presented in Fig. 5.2a. Dotted lines indicate spectral features explained in detail in the text. sp = spot, N-LDH = neo-formed Ni-Al LDH (14), LDH = synthetic Ni-Al LDH (Ni:Al, 2:1 (12)), Ni-Phyl = Ni-phyllosilicate (13).

The corresponding Fourier transforms (FT) of the k^3-weighted μ-EXAFS spectra are shown in Fig. 5.2b. The position of the first and second peak of the FT and the shape of the imaginary part reveals strong similarities to the Ni-Al LDH compounds. It should be noted that the amplitude of the second peak in the experimental spectra of the single spots is clearly reduced ((5,15); see discussion below). This observation also applies for the Ni-Al LDH reference compounds, but not for the other references. This finding corroborates the presence of Ni-Al LDH at the single spots in both Ni-doped HCP samples. This further indicates that Ni-Al LDH already forms in the early stage of the hydration process and that it persist in HCP hydrated for one year.

Table 5.1. Structural information obtained from selected μ-EXAFS Ni K-edge data analysis. (Spots are indicated in Fig. 5.1)

Samples	Ni-O			Ni-Ni			Ni-Si			ΔE_0	% Res
	CN	R (Å)	σ^2 (Å)	CN	R (Å)	σ^2 (Å)	CN	R (Å)	σ^2 (Å)		
References											
Ni-Phyllosilicate [a]	5.1	2.04	0.006	3.5	3.07	0.008[d]	3.7	3.3	0.008[d]	0.3	3.0
β-Ni(OH)$_2$	5.6	2.06	0.005	5.6	3.13	0.005[e]				-0.6	3.0
α-Ni(OH)$_2$	5.2	2.03	0.005	4.9	3.09	0.005[e]				3.0	4.4
Ni-Al LDH (LDH)	6.0	2.05	0.006	2.5	3.06	0.005[e]				1.1	4.5
Neo-formed Ni-Al LDH [b] (N-LDH)	5.7	2.04	0.004	3.9	3.07	0.005[e]				0.3	3.8
Cement samples											
Ni_cem_6 hrs (bulk-EXAFS)[c]	4.9	2.03	0.004	3.8	3.08	0.005[e]				-2.4	4.3
spot 1	5.8	2.05	0.006	3.3	3.08	0.005[e]				0.4	4.4
spot 2	5.6	2.05	0.006	3.0	3.10	0.005[e]				1.3	6.1
spot 3	7.3	2.06	0.009	3.0	3.10	0.005[e]				1.5	5.7
Ni_cem_1 y (bulk-EXAFS)[c]	6.5	2.04	0.006	3.0	3.09	0.005[e]				-1.8	4.1
spot 4	6.2	2.05	0.006	3.0	3.08	0.005[e]				-0.1	3.0
spot 5	5.5	2.05	0.005	3.3	3.08	0.005[e]				-0.2	5.3

[a] Scheidegger et al., 1997, [b] Dähn et al. 2002, [c] Vespa et al. 2006, [d] correlated parameters and [e] fix parameters during fitting procedures.
R, CN, σ^2 ΔE_0 stand for interatomic distances, coordination numbers, Debye-Waller factors and inner potential corrections.
Estimated error: R$_{(Ni-O)}$ ±0.02 Å, CN$_{(Ni-O)}$ ±20%, R$_{(Ni-Ni)}$ ±0.02 Å, CN$_{(Ni-Ni)}$ ±20%
% Res: deviation between experimental data and fit given by the relative residual in percent.
N= number of data points, Yexp and Ytheo: experimental and theoretical data points, respectively.

$$\% Res = \frac{\sum_{i=1}^{N}|y_{exp}(i) - y_{theo}(i)|}{\sum_{i=1}^{N} y_{exp}(i)} *100$$

The structural parameters derived from multi-shell analysis (R+ΔR range=0.8-3.5 Å) are summarized in Table 5.1. The first coordination shell was fitted with Ni-O backscattering pairs. The second coordination shell was fitted solely using Ni-Ni pairs, because the discrimination of Ni-Ni and Ni-Al backscattering pairs in Ni-Al LDH is problematic (3,4). To be able to compare the coordination numbers (CN) of the Ni-Ni backscattering pairs (CN$_{Ni-Ni}$) of all samples and references, the Debye-Waller factor was set to 0.005 Å^2. This value was determined from the fitting of the β-Ni(OH)$_2$ spectrum. Data analysis reveals similar CN and interatomic distances (R) for all spots. The first FT peak corresponds to an octahedral coordination of Ni with ~6 oxygen atoms at 2.05-2.06 Å. The second FT peak reveals strongly reduced CN_{Ni-Ni} (~3) compared to α-Ni(OH)$_2$ (~5) (16) and β-Ni(OH)$_2$ (~6) (17). The CN_{Ni-Ni} of the experimental spectra are comparable to those determined for Ni-Al LDH. The CN_{Ni-Ni} is reduced as Ni is partly substituted by Al in Ni-Al LDH. This causes a significant destructive interference between Ni and Al EXAFS contributions, which results in an amplitude cancellation of the Ni and Al shells (5,15). Although the CN_{Ni-Ni} of the single spots of the Ni-doped HCP sample and Ni-Al LDH agree very well, the overall Ni-Ni distances, especially of spot 2 and 3 (3.10 Å), are longer than those in Ni-Al LDH (3.06-3.07 Å). This finding suggests that, in addition to Ni-Al LDH, other Ni-containing phases are present. The longer R_{Ni-Ni} is attributed to the presence of β-Ni(OH)$_2$ impurities (R_{Ni-Ni}=3.13 Å) (5). Based on thermodynamic calculations, β-Ni(OH)$_2$ is expected to precipitate in the Ni-doped HCP samples, since at high pH the system is oversaturated with respect to this phase (18). The findings from the µ-EXAFS investigations indicate that at single Ni spots mainly Ni-Al LDH is formed, independent of the hydration time, and that only at some Ni spots small amounts of β-Ni(OH)$_2$ is formed. Linear combination (LC) of the experimental spectra of the single spots fitted with the reference compounds (α-Ni(OH)$_2$, β-Ni(OH)$_2$, Ni-Al LDH) indicate that at all single spots Ni-Al LDH is, indeed, the predominant phase forming, together with Ni-hydroxides (~40-50%). The Ni-hydroxides phase predominantly observed in the cement matrix was α-Ni(OH)$_2$, regardless of the hydration time. Additionally, β-Ni(OH)$_2$ was found at single spots (spots 2 and 3) in the HCP sample hydrated for six hours.

The results from the micro-scale study can be compared with those from earlier macro-scale investigation (5). The macro-scale (bulk-EXAFS) investigations explicitly revealed the formation of a mixture of Ni-Al LDH, α-Ni(OH)$_2$ and β-Ni(OH)$_2$ (5), for all hydration times. Furthermore, this earlier study demonstrated that, the composition of the mixture varied with the hydration time. In particular, it showed that the amount of Ni-Al LDH was found to increase with increasing hydration time, whereas α-Ni(OH)$_2$ decreased and β-Ni(OH)$_2$ remained constant. The µ-EXAFS experimental results presented in this study show that at single Ni-rich spots the same Ni phases precipitate, that is, predominantly Ni-Al LDH and some Ni-hydroxides. In the earlier study of Vespa et al. (5), the predominant Ni-hydroxide phase formed upon short hydration time was α-Ni(OH)$_2$, whereas β-Ni(OH)$_2$ was found to be the dominant Ni-hydroxide after longer hydration time. The micro-scale results appear to contradict this finding, since at short and long hydration time the predominant Ni-hydroxide phase observed, was α-Ni(OH)$_2$ This indicates that, µ-EXAFS measurements on a few spots, may not necessarily be representative to the whole matrix. Therefore, it is essential to complement micro-scale with macro-scale investigations to gain the full picture.

Based on both bulk- and μ-EXAFS results the following structural model for the Ni-Al LDH phase (Fig. 5.3) is proposed. Ni-Al layered double hydroxides are hydrotalcite-type phases commonly expressed as $[M^{II}_{1-x}M^{III}_{x}(OH)_2]^{x+}(A^{n-})_{x/n}\cdot yH_2O$. The M^{II} position can be filled with several bivalent metal cations (e.g., Mg, Mn, Fe, Co, Ni, Zn), the M^{III} position with trivalent metal cations (e.g., Al, Cr, Fe) and the A^{n-} position with different anions such as CO_3^{2-}, NO_3^-, Cl^-, SO_4^{2-}. In this study we propose that, the M^{II} and M^{III}, which lie at the centre of the octahedra, are partially filled with Ni and Al.

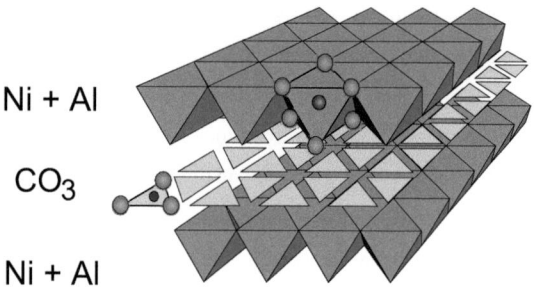

Fig. 5.3 Structural model of Ni-Al LDH. The centre of the octahedra are filled with Ni and Al. The interlayer position can be filled with different anions such as CO_3^{2-}, NO_3^-, Cl^-, SO_4^{2-}.

5.4 References

(1) Chapman, N.; McCombie, C. *Principles and standards for the disposal of long-lived radioactive wastes*; First ed.; Elsevier Science, Ltd.: Oxford, **2003**.

(2) Glasser, F. P. Chemistry of cement-solidified waste forms. . In: *Chemistry and microstructure of solidified waste forms*; Spence, R. D., Ed.; Lewis Publishers: Boca Raton, **1993**.

(3) Scheidegger, A. M.; Wieland, E.; Dähn, R.; Spieler, P. Spectroscopic evidence for the formation of layered Ni-Al double hydroxides in cement. *Environmental Science and Technology* **2000**, *34*, 4545-4548.

(4) Scheidegger, A. M.; Wieland, E.; Scheinost, A. C.; Dähn, R.; Tits, J.; Spieler, P. Ni phases formed in cement and cement systems under highly alkaline conditions: An XAFS study. *Journal of Synchrotron Radiation* **2001**, *8*, 916-918.

(5) Vespa, M.; Dähn, R.; Grolimund, D.; Wieland, E.; Scheidegger, A. M. Spectroscopic investigation of Ni speciation in hardened cement paste. *Environmental Science and Technology* **2006**, *40*, 2275-2282.

(6) Vespa, M.; Dähn, R.; Gallucci, E.; Grolimund, D.; Wieland, E.; Scheidegger, A. M. Micro-scale investigation of Ni uptake by cement using a combination of scanning electron microscopy and synchrotron-based techniques. *Environmental Science and Technology* **2006**, in press.

(7) Döhring, L.; Görlich, W.; Rüttener, S.; Schwerzmann, R. Herstellung von homogenen Zementsteinen mit hoher hydraulischer Permeabilität. *Nagra NIB 94-29* **1994**.

(8) Lothenbach, B.; Wieland, E. A thermodynamic approach to the hydration of sulphate-resisting portland cement. *Waste Management* **2006**, *26*, 706-719.

(9) Marcus, M.; MacDowell, A. A.; Celestre, R.; Manceau, A.; Miller, T.; Padmore, H. A.; Sublett, R. E. Beamline 10.3.2 at ALS: A hard- X-ray microprobe for environmental and material sciences. *Journal of Synchrotron Radiation* **2004**, *11*, 239-247.

(10) Ressler, T. WinXAS: A program for X-ray absorption spectroscopy data analysis under MS-Windows. *Journal of Synchrotron Radiation* **1998**, *5* (2), 118-122.

(11) Rehr, J. J.; Mustre de Leon, J.; Zabinsky, S. I.; Albers, R. C. Theoretical X-ray absorption fine structure standards. *Journal of the American Chemical Society* **1991**, *113*, 5135-5140.

(12) Johnson, C. A.; Glasser, F. P. Hydrotalcite-like minerals $(M_2Al(OH)_6(CO_3)_{0.5} \cdot XH_2O$, where M= Mg, Zn, Co, Ni) in the environment: synthesis, characterization and thermodynamic stability. *Clays and Clay Minerals* **2003**, *51*, 1-8.

(13) Dähn, R.; Scheidegger, A. M.; Manceau, A.; Schlegel, M. L.; Baeyens, B.; Bradbury, M. H.; Morales, M. Neoformation of Ni phyllosilicate upon Ni uptake on montmorillonite: A kinetics study by powder and polarized extended X-ray absorption fine structure spectroscopy. *Geochimica et Cosmochimica Acta* **2002**, *66* (13), 2335-2347.

(14) Scheidegger, A. M.; Lamble, G. M.; Sparks, D. L. The kinetics of nickel sorption on phyrophyllite as monitored by X-ray absorption fine structure (XAFS) spectroscopy. *Journal de Physique IV France* **1997**, *7* (C2), 773-775.

(15) Scheinost, A. C.; Sparks, D. L. Formation of layered single- and double-metal hydroxide precipitates at the mineral/water interface: A multiple-scattering XAFS analysis. *Journal of Colloid and Interface Science* **2000**, *223*, 1-12.

(16) Bode, H.; Dehmelt, K.; Witte, J. Zur Kenntnis der Nickelhydroxidelektrode-I. Über das Nickel (II)-Hydroxidehydrat. *Electrochimica Acta* **1966**, *11*, 1079-1087.

(17) Mansour, A. N.; Melendres, C. A. Analysis of X-ray absorption spectra of some nickel oxycompounds using theoretical standards. *Journal of Physical Chemistry A* **1998**, *102*, 65-81.

(18) Hummel, W.; Curti, E. Nickel aqueous speciation and solubility at ambient conditions: a thermodynamic elegy. *Monatshefte für Chemie* **2003**, *134*, 941-973.

CHAPTER 6

CO SPECIATION IN HARDENED CEMENT PASTE: A MACRO- AND MICRO-SPECTROSCOPIC INVESTIGATION

Abstract

Cement-based materials play an important role in multi-barrier concepts developed worldwide for the safe disposal of hazardous and radioactive wastes. Cement is used to condition and stabilize the waste materials and to construct the engineered barrier systems (container, backfill and liner materials) of repositories for radioactive waste. In this study, Co uptake by hardened cement paste (HCP) has been investigated with the aim of improving our understanding of the immobilization process of heavy metals in cement on the molecular level. X-ray-absorption spectroscopy (XAS) on powder material (bulk-XAS) was used to determine the local environment of Co in cement systems. Bulk-XAS investigations were complemented with micro-beam investigations to probe the inherent micro-scale heterogeneity of cement by using µ-X-ray-fluorescence (µ-XRF) and µ-XAS. µ-XRF was used to gain information on the spatial heterogeneity of Co distribution, whereas µ-XAS was employed to determine the speciation and oxidation state of Co on the micro-scale. The Co-doped HCP samples were prepared under normal atmosphere, to simulate similar conditions as for waste packages. To investigate the role of oxygen further samples were prepared in the absence of oxygen. The unhydrated cement powder was mixed by adding a $Co(NO_3)_2$ solution at a water/cement ratio of 0.4. Subsequently the samples were hydrated for different time-scales from 1 hour up to 1 year. The study showed that for the samples prepared in air Co(II) is oxidized to Co(III) after 1 hour of hydration time. Moreover, the relative amount of Co(III) increases with increasing hydration time. The study further revealed that Co(II) is predominately incorporated into newly formed Co-hydroxide-like phase and/or Co-phyllosilicates, whereas Co(III) tends to be incorporated into CoOOH-like phase and/or Co-phyllomanganates. In contrast to samples prepared in air, XAS experiments with samples prepared in the absence of oxygen revealed solely the presence of Co(II). This finding indicates that oxygen plays an important role for Co oxidation in cement. Furthermore the study suggests that Co(III) species or Co(III)-containing phases should be taken into account for an overall assessment of the Co release from Co-containing cement-stabilized waste under oxidizing conditions.

6.1 Introduction

Assuring safe disposal and long-term storage of hazardous and radioactive wastes represents a primary environmental task of industrial societies. The long-term disposal of the hazardous waste is associated with landfilling of cement-stabilized waste (e.g. 1), whereas deep geological disposal is foreseen for some categories of radioactive waste conditioned in cementitious materials (2). For example, more than 90 wt% of the near-field material of the planned Swiss geological repository for intermediate-level waste consists of hardened cement paste (HCP) and cementitious backfill materials. HCP is used to condition radioactive waste (solidification and stabilization). For this reason, an understanding of the immobilization processes in hydrating cement is essential to asses the behaviour of contaminants in cementitious waste matrices. From a chemical standpoint, HCP is a very heterogeneous material with discrete particles in the nano to micrometer size range. The material consists of mainly calcium (aluminum) silicate hydrates (C-(A)-S-H), portlandite (calcium hydroxide), calcium aluminates and ferrites. The immobilization potential of HCP originates from its selective binding properties for metal cations and anions (e.g., 3). Thus, it appears that immobilization processes in cement systems are highly specific with respect to the mineral components and mechanisms involved.

Co is an important contaminant in hazardous waste materials generated in medical applications and industrial processes. For example, Co^{60} is an important gamma-ray source, and is extensively used as a tracer and radiotherapeutic agent (4). Co radioisotopes may also be associated with irradiated metallic components from nuclear power plants, and can be, therefore, present in cement-stabilized radioactive waste. In this case, a mechanistic understanding of Co immobilization is of major importance for predicting the long-term fate of Co in the cementitious near-field of repositories for radioactive waste. In connection with the disposal of hazardous industrial and municipal waste, molecular level information on the speciation of Co will allow a more detailed assessment of the leachability of the heavy metal, e.g. from landfills and contaminated soils into aquifers, to be made.

While Co speciation has been widely investigated in natural minerals, in clays or soils (e.g., 5,6,7), in catalysis reactions or alkaline solutions (e.g., 8,9), only very few studies have addressed uptake mechanisms of Co by cementitous materials. Masse et al. (10) observed the formation of Co(II)-silicate phases and Co(II)/(III)-oxide (Co_3O_4) in heat-treated cements. Komarneni et al. (11) showed that Co(II) almost completely replaces Ca(II) in the structure of crystalline C-S-H-phases. Scheidegger et al. (12) identified both oxidation states of Co, i.e.,

Co(II) and Co(III), within a microbial corroded Co-containing cementitious waste by using μ-XRF and μ-XAS techniques.

The objective of the present study was to investigate the Co speciation during cement hydration. The hydration process was started by adding Co(II) salt solution to the unhydrated cement. The procedures used in the experiments should simulate real conditions under which cementitious waste packages are usually produced. In particular, hydration was carried out under normal atmospheric conditions. To elucidate the influence of oxygen on the Co speciation, Co-doped HCP samples were also prepared in the absence of oxygen. X-ray-absorption spectroscopy (XAS) on Co-doped powder material (bulk-XAS) was used to determine the Co speciation and gain a molecular level understanding of the immobilization processes. To further investigate the elemental distribution, in-situ speciation and oxidation state of Co on the micro-scale, μ-XRF and μ-XAS were employed. The multi-technique approach allows the structural information gained from micro-scale and macro-scale investigations to be compared with each other and the predominant Co species formed in the cement matrix to be identified.

6.2 Materials and methods

6.2.1 Sample preparation

The cement samples were prepared from a commercial sulphate-resisting Portland cement (CEM I 52.5 N HTS, Lafarge, France) used to condition radioactive waste in Switzerland. Co(II)-doped HCP samples were prepared by mixing $Co(NO_3)_2$ solution with unhydrated cement. The metal salt was dissolved in deionized water to obtain a stock solution with a concentration of 0.3 mol/L (pH=5-6). The solution was mixed with the unhydrated cement using a water/cement (w/c) ratio of 0.4 and a standard procedure (13). The final metal concentration of the pastes was 5000 mg/kg dry HCP. Note that the preparation of these samples occurred under normal atmosphere. The cement pastes were filled into Plexiglas moulds, which were closed with polyethylene lids, and hydrated between 1 hours and ~1 year in a water-saturated atmosphere in closed containers. For short hydration times of 1 hour, the slurry was filtered (0.2 μm pore size) to separate the solid from the free water. The solid materials was washed with acetone for 15 minutes to stop the hydration process (14), filtered and dried in a glovebox under controlled N_2 atmosphere (CO_2, O_2<2 ppm, T=20±3 °C). A Co(II)-doped HCP sample (tot-O_2) hydrated for 1 hour was completely prepared in the glovebox. Special measures were taken to reduce the oxygen associated with the cement powder and the water used for the preparation of the stock solution, e.g., by purging the water with nitrogen for 24 hours and repeatedly evacuating the cement powder under high vacuum.

Each cylinder was cut into several slices of ~1 cm thickness and dried in the glovebox. Some slices were crushed to obtain size fractions <100 μm using a tungsten/carbide pebble mill. For bulk-XAS measurements the powder material was filled into Plexiglas holders and sealed with Kapton tape. Other slices were impregnated and polished for the preparation of thin sections, which were employed for μ-XRF and μ-XAS investigations.

6.2.2 XAS and μ-XRF data collection and reduction

Bulk-XAS spectra at the Co K-edge were collected at the Swiss Norwegian Beam Line (SNBL) and the Dutch Belgium Beamline (DUBBLE) at the European Synchrotron Radiation Facility (ESRF) in Grenoble, France. The measurements were collected at room temperature in transmission (ionization chambers) and in fluorescence mode (SNBL: Lytle detector; DUBBLE: 9 channel monolithic Ge-solid state detector).

μ-XRF and μ-XAS data were collected on beamline 10.3.2 at the Advanced Light Source (ALS), Berkeley, USA (15). The measurements were collected at room temperature in fluorescence mode using a 7-element Ge-solid state detector with a beam size of ~5x5 μm^2. The μ-XRF maps were obtained by scanning the sample under the monochromatic beam at the energy of 10000 eV with a beam size of ~5x5 μm^2.

All beamlines are equipped with a Si(111) crystal monochromator. The monochromator angle was always calibrated by assigning the energy of 7709 eV to the first inflection point of the K-absorption edge spectrum of Co metal foil.

The μ-XRF maps were processed using the Labview software package at beamline 10.3.2 (15) and MATLAB. μ-XAS and bulk-XAS data reduction was performed using the WinXAS 3.1 software package following standard procedures (16). All spectra were normalized by fitting a first-degree polynomial to the pre-edge and a third-degree polynomial to the post-edge regions. The energy was converted to photoelectron wave vector units (Å^{-1}) by assigning the origin E_0 to the first inflection point of the absorption edge. Radial Structure Functions (RSF) were obtained by Fourier transforming the k^3-weighted $\chi(k)$ functions between 3.2 and 10.9 Å^{-1} with a Bessel window function using a smoothing parameter of 4. Multishell fits were performed in real space across the range of the first two shells. Theoretical scattering paths for the fit were calculated using FEFF 8.20 (17) and the structure of Co(OH)$_2$ and CoOOH as a reference. The amplitude reduction factor (S_0^2) was set to 0.85

(18). Errors on the structural parameters were estimated from the analysis of a series of reference compounds (Co(OH)$_2$, CoOOH). Several reference spectra Co(OH)$_2$, synthetic Co-Al LDH (Co:Al, 2:1; Co$_2$Al(OH)$_6$(CO$_3$)$_{1/2}$ (19)), Co-phyllosilicate (e.g., Co-kerolite (20)), CoOOH, Co-phyllomanganate (e.g., Co-asbolane (21), Co-buserite (22)), were used to identify the Co speciation in the cement matrix.

6.3 Results and Discussion
6.3.1 Influence of hydration time on the Co oxidation state

Thermodynamic calculation using the available thermodynamic data (23,24) indicate that, in an initial phase of the hydration process, i.e., when adding the concentrated Co solution (5000 mg/kg) to the cement, the system is strongly oversaturated with respect to Co(OH)$_2$. Thus, the precipitation of Co(OH)$_2$ was anticipated to occur in the system. Moreover, the Eh of the cement system is not well buffered. The condition in cement matrices can vary from slightly oxidizing to slightly reducing, depending on the composition of the cementitious material (25). Thus, under the given conditions and at pH>12.5, typical for cement matrices, the formation of Co(III)-containing phases during hydration together with Co(OH)$_2$ could not be excluded (Figure S1, Supporting Information).

Fig. 6.1 shows the normalized, background-subtracted and k^3-weighted bulk- and µ-extended X-ray absorption fine structure (EXAFS) spectra (Fig. 6.1a), the corresponding radial structure functions (RSFs) (Fig. 6.1b) and the Fourier-backstransformed (FT^{-1}, $\Delta R=0.8$-7 Å, Fig. 6.1c) spectra of all of the Co(II)-doped cement samples with hydration times of 1 hour, 3 days and 1 year, and relevant reference spectra. Further, Fig.6.2 shows the corresponding k^3-weighted EXAFS functions for the Fourier-backtransformed spectra of the 1st (Fig.6.2a) and 2nd shell (Fig.6.2b) obtained from Fig.6.1b. Note that all samples were prepared in air except the sample denoted as 1hour-tot-O$_2$. In the following discussion, only the data of the Co(II)-doped powder materials (bulk-EXAFS) prepared in air will be compared.

Fig.6.1a and c reveal that the spectrum of the sample hydrated for 1 hour shows similarities with those of the Co(II) reference compounds. The shape of the oscillation at ~4 Å^{-1} shows similarities to that of Co-hydroxide. The oscillations at~6 Å^{-1} and between ~7-8.5 Å^{-1} further support similarities with the Co(II) spectra, especially to Co-hydroxide and/or Co-phyllosilicate (Co-kerolite). However, the corresponding RSF (Fig.6.1b) reveals a relatively broad peak for the first shell, suggesting that not pure Co(II) species formed in the cement matrix after 1 hour hydration time. The presence of both Co(II) and Co(III) species is clearly indicated in the corresponding FT^{-1} of the first shell (Fig.6.2a), which shows a slight frequency shift to higher k-values and a reduced amplitude above ~8 $Å^{-1}$ compared to the Co(II) references. The reduced amplitude is attributed to the destructive interference of Co(II)-O and Co(III)-O backscattering contributions, causing an amplitude cancellation. Note, however, that the RSF (Fig.6.1b) and FT^{-1} (Fig.6.2b) of the second shell are comparable to Co(II)-hydroxide and/or Co(II)-phyllosilicate. The above findings indicate that, after 1 hour hydration time, predominantly Co(II) species are present (most likely Co-hydroxide and/or Co-phyllosilicate) in the cement sample prepared in air. Nevertheless, the results clearly show that small amounts of Co(III) had formed.

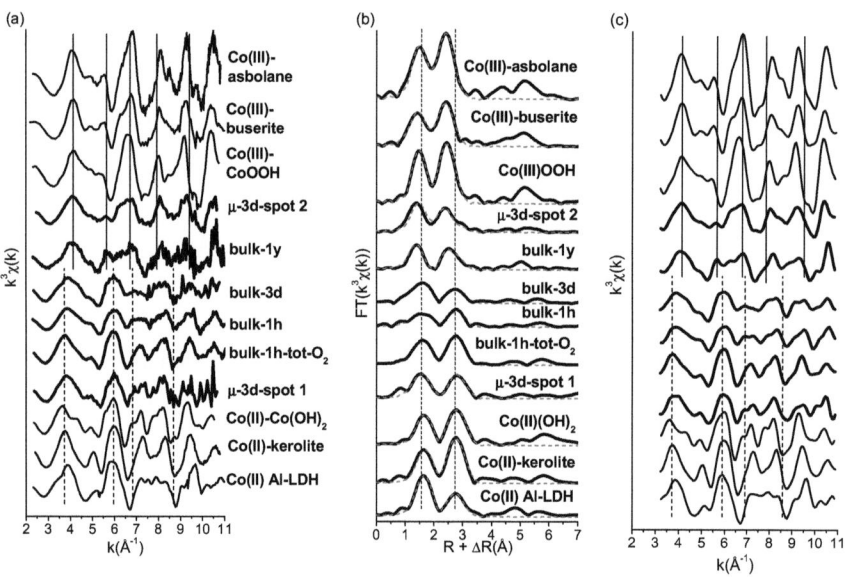

Fig. 6.1 Experimental spectra of Co-reference compounds and Co(II)-doped HCP (5000 mg/kg metal loading) of powdered samples (bulk) hydrated between 1 hour and 1 year and single micro-spot spectra of a sample hydrated for 3 days (spot 1 and 2). A further Co(II)-doped HCP spectrum hydrated for 1 hour and prepared in the absence of oxygen (bulk-1h-tot-O_2) is shown. (a) k^3-weighted, normalized, background-subtracted EXAFS spectra; (b) Radial Structure Functions obtained from the Fourier transform of the EXAFS spectra presented in Fig.6.1a; (c) k^3-weighted EXAFS function for the Fourier-backtransformed spectra obtained from Fig. 6.1b (range: R+ΔR=0.8-7 Å). Dashed lines indicate spectral

features typical of Co(II) species; solid lines indicate spectral features typical of Co(III) species. Co-Al LDH (Co:Al, 2:1; $Co_2Al(OH)_6(CO_3)_{1/2}$ (19); Co-kerolite (20); Co-asbolane (21); Co-buserite (22) h = hour; d = days; y = year.

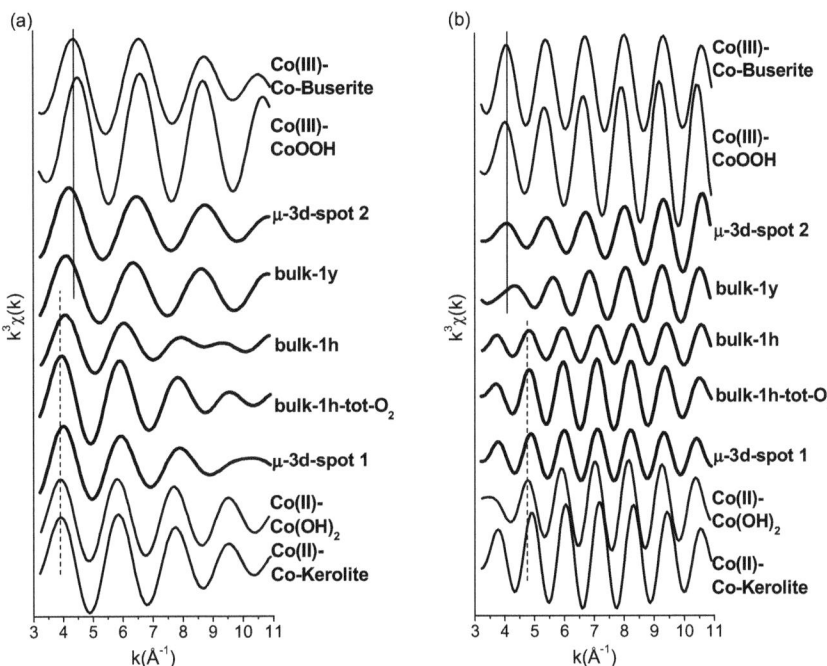

Fig.6.2. k^3-weighted EXAFS function for the Fourier-backtransformed spectra of the 1st (a) and 2nd shell (b) obtained from Fig.6.1b. The dashed line indicates the phase position for the Co(II) species, whereas the solid line indicates the phase position for the Co(III) species.

The bulk-EXAFS spectrum of the sample hydrated for 1 year reveals that the oscillation at ~4 Å^{-1} is different from that in the 1-hour sample, but very similar to those of the Co(III) reference compounds, that is to CoOOH and/or Co-phyllomanganates (e.g. Co-buserite, Co-asbolane; Fig.6.1a and c). Additionally the splitting of the oscillation at ~6 Å^{-1} indicates the predominant formation of Co(III) species. Further, the beat pattern observed between ~8-9.5 Å^{-1} also show strong similarities to the Co(III) species. Again, the formation of Co(III) species is better visualized in the corresponding FT^{-1} of the first (Fig.6.2a) and second shell (Fig.6.2b). For both shells the FT^{-1} reveals strong similarities to the Co(III) reference compounds. The corresponding RSF (Fig.6.1b) of the first shell and second shell further show a clear shift to shorter distances with peak positions ($R+\Delta R^{\text{1st-shell}}=\sim 1.40$ Å; $R+\Delta R^{\text{2nd-shell}}=\sim 2.51$ Å) comparable to Co(III) reference spectra ($R+\Delta R^{\text{1st-shell}}=\sim 1.40$ Å; $R+\Delta R^{\text{2nd-shell}}=\sim 2.42$ Å). Although both distances are comparable to Co(III) compounds, the peak position of the second shell is longer than any Co(III) species ($R+\Delta R=\sim 2.42$ Å), indicating that small amounts of Co(II) are still present in the 1 year hydrated cement sample.

The bulk-EXAFS spectrum of the Co-doped cement sample hydrated for 3 days appears to be a combination of the spectra of the two other samples, further supporting the co-existence of Co(II) and Co(III) species in the samples. This is particular visible in the bulk-EXAFS spectrum and FT^{-1} (Fig.6.1a and c) between ~7-8.5 $Å^{-1}$.

Linear combination fits (LC) were carried out to estimate the ratio of Co(II) and Co(III) species form in the Co(II)-doped cement samples. LC of the bulk-EXAFS spectra revealed that ~70% Co(II) and ~30% Co(III) are present in the cement matrix after 1 hour hydration. The ratio of Co(II) to Co(III) species significantly changes with time. After 1 year hydration ~40% of Co remained as Co(II), whereas ~60% of Co existed in the form of Co(III) species. The above findings show that Co(II) is rapidly oxidized to Co(III), i.e., within the first hours of cement hydration in normal atmosphere, and that the oxidation process continuous with ongoing hydration. As a consequence, Co(III) species dominate in the cement matrix after a hydration time of 1 year with small amounts of Co(II) species.

6.3.2 Distribution and speciation of Co on the micro-scale

The bulk-EXAFS investigation of the Co(II)-doped HCP samples revealed the presence of a mixture of Co(II) and Co(III) species in the cement matrix. To further investigate whether these species can be individually identified on the micro-scale spatially-resolved μ-XRF and μ-XAS investigation were performed. μ-XRF was employed in a first step to localize the Co enriched areas to be studied by μ-XAS and to gain information on possible correlations of Co with other elements.

Fig.6.3 shows elemental distribution maps of Co, Ca and Fe for two different regions of the Co(II)-doped HCP sample hydrated for 3 days. Ca is the main constituent of the cement matrix, whereas Fe(III) is a possible oxidizing agent (see following section) promoting Co(II) oxidation. Fig.6.3 illustrates the heterogeneous distributions of all elements as mapped by μ-

XRF. The spatial distribution of all elements reveals higher (red for Co, white for Ca, yellow for Fe) and lower (blue for Co, green for Ca, dark red for Fe) concentrated areas. The areas with high Ca and Fe concentrations indicate the presence of clinker minerals (non-hydrated cement). The less concentrated Ca and Fe areas correspond to hydrated cement phases (26). In both mapped regions (Fig.6.3a and b) an anti-correlation of Co with Ca and Fe was observed, suggesting that Co is neither associated with Ca- nor Fe-containing phases. Furthermore, μ-XRF investigations reveal that Co is enriched either in spot-like or ring-like structures. The ring-like structures form around several clinker minerals (regions with high Ca concentrations, Fig.6.3b) and some hydrated cement phases (low Ca concentrations, Fig.6.3b).

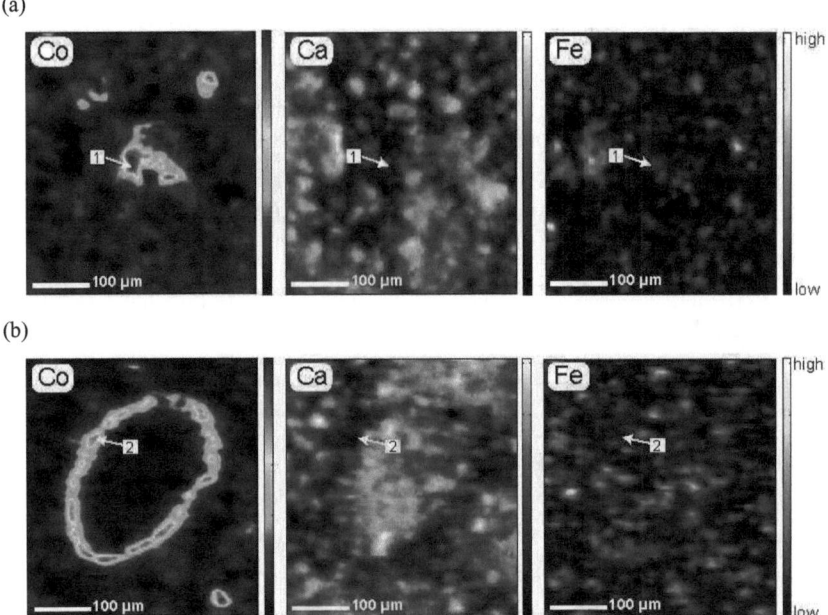

Fig.6.3. μ-XRF elemental distribution maps of Co, Ca and Fe for the Co(II)-doped HCP sample (5000 mg/kg metal loading) hydrated for 3 days (a) Co rich spot, (b) Co ring-like structure. The color bar represents relative concentrations in each sample. The μ-EXAFS data of spots 1 and 2 are shown in Fig.6.1.

In a next step, Co speciation and oxidation states were determined by μ-XAS at areas located on Co-enriched spot-like (e.g., spot 1 in Fig.6.3a) and ring-like structures (e.g., spot 2 in Fig.6.3b). The selected μ-EXAFS measurements collected on spot 1 and spot 2 are shown in Fig.6.1. The k^3-weighted μ-EXAFS spectrum of spot 1 (Fig.6.1a) reveals that, as in the case of the bulk measurements, the first oscillation at ~4 Å^{-1} corresponds to that of Co(II) compounds. Similarly, the small feature at ~5 Å^{-1} as well as the oscillations at ~6 Å^{-1} and between 7-8.5 Å^{-1}, previously observed in the bulk-EXFAS spectra and better visualized in the FT^{-1} (Fig.6.1c; ΔR=0.8-7 Å), are also comparable to Co(II) compounds. The corresponding RSF and FT^{-1} of the first and second shell are shown in Fig.6.1 and Fig.6.2, respectively. Although the figures show a small shift in the FT$^{1st\text{-}shell}$ peak position (R+ΔR=~1.51 Å, Fig.6.1b) and a frequency shift of the first (Fig.6.2a) and second shells (Fig.6.2b) in the FT^{-1} of the spot 1 spectrum compared to Co(II) reference spectra (R+ΔR=~1.65 Å), data analysis yields structural parameters (2.06 Å and 3.16 Å) that are similar to those of Co-hydroxide ($R_{Co\text{-}O}$=2.09 and $R_{Co\text{-}Co}$=3.16; Table S1, Supporting Information (20)). The above findings reveal that at spot 1 a Co(II) species has formed, predominantly Co-hydroxide and/or Co-phyllosilicate. Nevertheless, the spectrum collected at sport 1 is reproduced neither by a pure Co-hydroxide nor by a pure Co-phyllosilicate. This indicates that rather a mixture of different Co(II)-phase has formed or that the size of the various Co(II)-forming phases is smaller than the used beam size (5x5 μm^2), causing an average signal.

In contrast, the spectrum of spot 2 located on the ring-like structure reveals similarities to the spectra of Co(III) species. The oscillations of the k^3-weighted EXAFS spectrum (Fig.6.1a and c) as well as those of the FT^{-1} of the first (Fig.6.2a) and second shell (Fig.6.2b) are well in phase with the spectra of Co(III) compounds (CoOOH and/or Co-phyllomanganate). Furthermore, the small feature at ~5.5 Å^{-1}, the oscillations at ~7 Å^{-1} and between ~8-9.5 Å^{-1} further support the idea that Co(III) compounds dominate at this spot (Fig.6.1a and c). Finally, the RSF (Fig.6.1b) shows peak positions for the first (R+ΔR=~1.38 Å) and second shells (R+ΔR=~2.39 Å), which are comparable to those of Co(III) compounds (R+ΔR$^{1st\text{-}shell}$=~1.40 Å; R+ΔR$^{2nd\text{-}shell}$=~2.42 Å). Analysis of the EXAFS data collected on spot 2 results in typical Co(III)-O and Co(III)-Co/Mn distances ($R_{Co\text{-}O}$=~1.90 and $R_{Co\text{-}Co/Mn}$=~2.80-2.85; Table S1, Supporting Information (21,22). Note that the overall discrimination between CoOOH and Co-phyllomanganates is difficult to achieve because the backscattering pair contributions of Co and Mn result in similar EXAFS and FT functions (21,22). Note, however, that the Mn concentration is low (~200 mg/kg) compared to the Co concentration in the cement (~5000 mg/kg), which implies that, from a stochiometric point of view, the formation of significant amounts of Co(III)-containing phyllomanganates is less probable.

The above findings indicate that Co(II) and Co(III) phases form at separate micro spots within the Co(II)-doped cement matrix. While, the bulk-XAS measurements suggest the formation of a mixture of Co(II) and Co(III) phases, the µ-EXAFS experiments clearly show that these species form in distinct areas of the cement matrix. Furthermore, it appears that Co(II) precipitates mainly as Co-hydroxide and/or Co-phyllosilicate, whereas Co(III) forms CoOOH and/or Co-phyllomanganate.

6.3.3 The role of oxygen in Co(II) oxidation

Several potentially oxidizing agents, i.e. O_2, Fe(III) and Mn(IV), were considered in our approach towards a better understanding of Co(II) oxidation in the Co(II)-doped HCP samples. In the first step the Fe(II)/Fe(III) redox couple was excluded as X-ray absorption near edge spectroscopy (XANES) spectra collected at the Fe K-edge in reactive zones of the cement matrix with significant Co(III) accumulation provided no evidence for Fe(II) formation (Figure S2, Supporting Information).

Further, the Mn(II)/Mn(IV) redox couple was excluded as electron paramagnetic resonance (EPR) measurements showed that the Mn(II) concentration in the cement matrix decreased with increasing hydration time. Note that both the concentrations of Co(II) and Mn(II) were found to decrease with increasing hydration time (Figure S3, Supporting Information), indicating that for both elements the higher oxidations state are promoted with time. The above finding suggests that the same oxidizing agent might be responsible for the oxidation of the bivalent metal cations in the cement matrix.

The role of O_2 in controlling the oxidation state of Co in the hydrating cement was further investigated. For this reason, Co(II)-doped HCP samples were prepared in the glovebox under nitrogen atmosphere in the absence of oxygen. The corresponding bulk-EXAFS spectrum for the sample reacted for 1 hour (bulk-1hour-tot-O_2) is shown in Fig.6.1. Fig.6.1a and c reveal for the bulk-1hour-tot-O_2 typical features observed for Co(II) species, i.e., an elongated upward oscillation ending in a sharp tip at ~3.7 Å, the small spectral feature at ~5 Å^{-1}, and the oscillation at ~6 Å^{-1}. Further, the spectrum shows two oscillations between ~7-8.5 Å^{-1} comparable to Co-phyllosilicate. The corresponding FT (Fig.6.1b) shows that the amplitude of the first and second peak of the bulk-1htot-O_2 spectrum show strong similarities to Co-hydroxide and Co-phyllosilicate. This finding is further supported by the FT^{-1} of the first (Fig.6.2a) and second shell (Fig.6.2b).

Data analysis finally revealed distances of ~2.07 Å for Co-O and of ~3.14 Å for Co-Co backscattering pairs, typical of Co(II) species (Table S1, Supporting Information (20)). The results for this Co(II)-doped HCP sample prepared in the absence of oxygen revealed the predominant formation of Co(II)-containing phases (Co-hydroxide and/or Co-phyllosilicate).

6.3.4 Implications of Co(II) oxidation in cementitious systems

This study demonstrates that the extent of Co(II) oxidation in cementitious matrices depends on the oxygen concentration present in the system. Oxygen is dissolved in the water and trapped in the cement powder. Finally, mixing water with cement in air to start the hydration process may further increase the oxygen concentration in the cementitious system. The results clearly showed that the oxidation of Co(II) to Co(III) in cementitious systems is fast, and that it is very difficult to avoid the Co(III) formation during cement hydration unless sample preparation is carried out under strictly anoxic conditions.

The above finding has implications for an overall assessment of the Co immobilization in cement-stabilized hazardous and radioactive wastes. Firstly, Co(III) is expected to be the dominant valence state in waste matrices that were produced under oxidizing conditions and stored in contact with air, e.g., Co-containing waste resulting from municipal and industrial processes stabilized with cement prior to landfilling. Secondly, Co(II) is expected to be the main valence state in repositories for radioactive wastes in which reducing conditions prevail in the long term (27). Nevertheless, the results of our investigations clearly show the difficulties that have to be overcome in conjunction with the determination of Co(II) retention in cementitious systems. An appropriate experimental setup for measuring the sorption of Co(II) onto cementitious materials or the solubility of Co(II)-containing solid phase in cementitious systems requires that the presence of oxygen is strictly avoided, and that the oxidation state of Co in these systems is carefully checked using spectroscopic techniques. For example, several studies in the past observed Co concentrations in cementitious systems that were lower than predicted based on the solubility of Co-hydroxides (e.g., 28). The present study suggests that Co(III) was the dominant oxidation state of Co in these systems. To further address the problem associated with the problematic of the proper interpretation of Co(II) sorption studies, bulk-EXAFS investigations on samples of Co(II)-sorbed onto HCP prepared in a glovebox under controlled N_2 atmosphere (CO_2, O_2<2 ppm) were carried out. The bulk-EXAFS data clearly show that Co(II)-sorbed onto HCP is oxidized to Co(III), thus forming Co(III)-containing phase (Figure S4, Supporting Information). This finding can be attributed to the fact that O_2 was not completely removed from the system, e.g., by purging the water for 24 hours with N_2 and repeatedly evacuating a chamber containing the HCP powder under high vacuum.

Supporting Information Available

Eh vs. pH diagram for Co, and further Co and Fe K-edge bulk/µ-XANES and EPR data.

6.4 References

(1) Schmidt, M.; Beckefeld, P.; Götz, R.; Kamsties, S.; Kretz, C.; Molitor, N.; Neck, U.; Vogel, P. *Reststoff-und Abfallverfestigung. Immobilisierung von Schadstoffen-Recycling-Verbesserung der Deponiefähigkeit*; Expert Verlag: Renningen-Malmheim, **1995**.

(2) Chapman, N.; McCombie, C. *Principles and standards for the disposal of long-lived radioactive wastes*; First ed.; Elsevier Science, Ltd.: Oxford, **2003**.

(3) Glasser, F. P. Chemistry of cement-solidified waste forms. . In: *Chemistry and microstructure of solidified waste forms*; Spence, R. D., Ed.; Lewis Publishers: Boca Raton, **1993**.

(4) Weast, R. C.; Astle, M. J. *CRC Handbook of chemistry and physics* Boca Raton, Florida, **1983**.

(5) Schlegel, M. L.; Charlet, L.; Manceau, A. Sorption of metal ions on clay minerals II. Mechanism of Co sorption on hectorite at high and low ionic strength and impact on the sorbent stability. *Journal of Colloid and Interface Science* **1999**, *220*, 392-405.

(6) Catalano, J. G.; Warner, J. A.; Brown Jr., G. E. Sorption and precipitation of Co(II) in Hanford sediments and alkaline aluminate solutions. *Applied Geochemistry* **2005**, *20*, 193-205.

(7) Zachara, J. M.; Cowen, C. E.; Resch, C. T. Sorption of divalent metals on calcite. *Geochimica et Cosmochimica Acta* **1991**, *55*, 1549-1562.

(8) Moen, A.; Nicholson, D. G.; Rønning, M.; Lamble, G. M.; Lee, J.-F.; Emerich, H. X-ray absorption spectroscopic study at the cobalt K-edge on the calcination and

reduction of the microporous cobalt silicoaluminophosphate catalyst CoSAPO-34. *Journal of the Chemical Society, Faraday Transaction* **1997**, *93* (22), 4071-4077.

(9) Totir, D.; Mo, Y.; Kim, S.; Antonio, M. R.; Scherson, D. A. In situ Co K-edge x-ray absorption fine structure of cobalt hydroxide film electrodes in alkaline solutions. *Journal of the Electrochemical Society* **2000**, *147* (12), 4594-4597.

(10) Masse, S.; Boch, P.; Vaissière, N. Trapping of nickel and cobalt in $CaNiSi_2O_6$ and $CaCoSi_2O_6$ diopside-like phases in heat-treated cement. *Journal of the European Ceramic Society* **1999**, *19*, 93-98.

(11) Komarneni, S.; Roy, R.; Roy, D. M. Pseudomorphism in Xonotlite and Tobermorite with Co^{2+} and Ni^{2+} exchange for Ca^{2+} at 25°C *Cement and Concrete Research* **1986**, *16*, 47-58.

(12) Scheidegger, A. M.; Grolimund, D.; Cheeseman, C. R.; Rogers, R. D. Micro-spectroscopic investigations of highly heterogeneous waste repository materials. *Journal of Geochemical Exploration* **2006**, *88*, 59-63.

(13) Döhring, L.; Görlich, W.; Rüttener, S.; Schwerzmann, R. Herstellung von homogenen Zementsteinen mit hoher hydraulischer Permeabilität. (Wettingen, Switzerland), unpublished report.

(14) Lothenbach, B.; Wieland, E. A thermodynamic approach to the hydration of sulphate-resisting portland cement. *Waste Management* **2006**, *26*, 706-719.

(15) Marcus, M.; MacDowell, A. A.; Celestre, R.; Manceau, A.; Miller, T.; Padmore, H. A.; Sublett, R. E. Beamline 10.3.2 at ALS: A hard- X-ray microprobe for environmental and material sciences. *Journal of Synchrotron Radiation* **2004**, *11*, 239-247.

(16) Ressler, T. WinXAS: A program for X-ray absorption spectroscopy data analysis under MS-Windows. *Journal of Synchrotron Radiation* **1998**, *5* (2), 118-122.

(17) Rehr, J. J.; Albers, R. C. Theoretical approaches to X-ray absorption fine structure. *Reviews of Modern Physics* **2000**, *72* (3), 621-653.

(18) O'Day, P. A.; Rehr, J. J.; Zabinsky, S. I.; Brown, J. G. E. Extended X-ray absorption fine structure (EXAFS) analysis of disorder and multiple-scattering in complex crystalline solids. *Journal of the American Chemical Society* **1994**, *116*, 2938-2949.

(19) Johnson, C. A.; Glasser, F. P. Hydrotalcite-like minerals $(M_2Al(OH)_6(CO_3)_{0.5} \cdot XH_2O$, where M= Mg, Zn, Co, Ni) in the environment: synthesis, characterization and thermodynamic stability. *Clays and Clay Minerals* **2003**, *51*, 1-8.

(20) Manceau, A.; Schlegel, M.; Nagy, K. L.; Charlet, L. Evidence for the formation of trioctahedral clay upon sorption of Co^{2+} on quartz. *Journal of Colloid and Interface Science* **1999**, *220*, 181-197.

(21) Manceau, A.; Llorca, S.; Galas, G. Crystal chemistry of cobalt and nickel in lithiophorite and asbolane from New Caledonia. *Geochimica et Cosmochimica Acta* **1987**, *51*, 105-113.

(22) Manceau, A.; Drits, V.; Silvester, E.; Bartoli, C.; Lanson, B. Structural mechanism of Co^{2+} oxidation by the phyllomanganate buserite. *American Mineralogist* **1997**, *82*, 1150-1175.

(23) Baes, C. F.; Mesmer, R. E. *The hydrolysis of cations*; John Wiley & Sons, Inc.: New York, **1976**.

(24) Hem, J. D.; Roberson, C. E.; Lind, C. J. Thermodynamic stability of CoOOH and its coprecipitation with manganese. *Geochimica et Cosmochimica Acta* **1985**, *49*, 801-810.

(25) Macphee, D. E.; Glasser, F. P. Immobilization science of cement systems. *MRS Bulletin* **1993**, *March* 66-71.

(26) Vespa, M.; Dähn, R.; Gallucci, E.; Grolimund, D.; Wieland, E.; Scheidegger, A. M. Micro-scale investigation of Ni uptake by cement using a combination of scanning electron microscopy and synchrotron-based techniques. *Environmental Science and Technology* **2006**, in press.

(27) Wieland, E.; Van Loon, L. R. Cementitious near-field sorption data base for performance assessment of an ILW repository in Opalinus Clay. *Nagra Technical Report NTB 02-20* **2002**.

(28) Baur, I.; Ludwig, C.; Johnson, C. A. Leaching behaviour of cement-stabilised incinarator ashes: A comparison of field and laboratory measurements. *Environmental Science and Technology* **2001**, *35* (13), 2817-2822.

SUPPORTING INFORMATION FOR CHAPTER 6
Co Speciation in Hardened Cement Paste:
A Macro- and Micro-spectroscopic Investigation

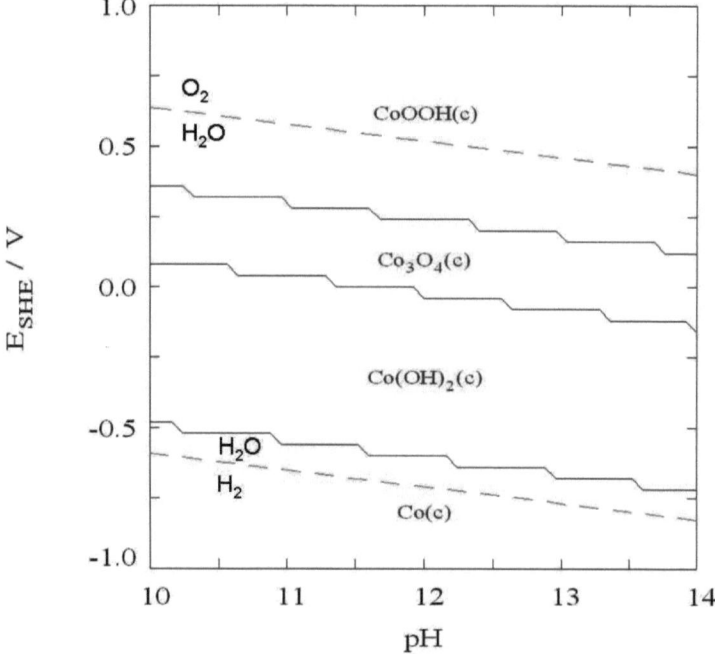

Figure S6.1. Eh vs. pH diagram showing the stability boundaries of Co(II) and Co(III) precipitates with Co^{2+} at 25C° 0.3 M Co^{2+}. The MEDUSA program was used for the calculation of the Eh vs. pH diagram. The thermodynamic data for the hydrolysis species were taken from (1) and for the solid phases of CoOOH and Co_3O_4 from (2). Co_3O_4 is not expected to precipitate in the investigated Co(II)-doped HCP matrix since this is a high temperature phase.

Table S6.1. Structural information obtained from selected bulk and μ-EXAFS Co K-edge data analysis with reference compounds (3-5) (spots are indicated in Fig. 4).

	1st shell						2nd shell							
	Co(II)-O			Co(III)-O			Co(II)-Co(II)			Co(III)-Co(III)/Mn				
Samples	CN	R (Å)	σ^2 (Å^2)	CN	R (Å)	σ^2 (Å^2)	CN	R (Å)	σ^2 (Å^2)	CN	R (Å)	σ^2 (Å^2)	ΔE_0 (eV)	%Res
References														
Co(OH)$_2$	4.1	2.09	0.005	-	-	-	5.0	3.16	0.007	-	-	-	-7.3	5.5
Co-Al LDH	6.5	2.09	0.006	-	-	-	2.3	3.09	0.006	-	-	-	0.3	3.0
Co-Kerolite[a]	4.4	2.09	0.005	-	-	-	7.2	3.13	0.008	-	-	-	-4.6	6.9
CoOOH	-	-	-	4.0	1.90	0.001	-	-	-	4.6	2.85	0.004	0.7	
Co-Buserite[b]	-	-	-	5.9	1.89	0.009	-	-	-	4.6	2.80	0.007	-1.8	27.6
Co-Asbolane[c]	-	-	-	5.2	1.89	0.004	-	-	-	7.2	2.80	0.007	-1.2	3.9
Cement samples														
μ_3d_spot 1	4.3	2.06	0.009	-	-	-	4.1	3.16	0.009[e]	-	-	-	6.2	17.4
μ_3d_spot 2	-	-	-	5.3	1.90	0.01	-	-	-	1.8	2.80	0.005[d]	-2.0	15.3
bulk_tot_O$_2$_1h	3.6	2.07	0.005[d]	-	-	-	3.1	3.14	0.005[d]	-	-	-	3.1	4.0

[a] Mancaeuet al., 1999, [b] Mancaeuet al., 1997, [c] Mancaeuet al., 1987, [d] fix parameters during fitting procedures and [e] correlated parameters
CN, R, σ^2, ΔE_0 stand for interatomic distance, coordination number, Debye-Waller factor and inner potential correction.
Estimated errors: R$_{(Co-O)}$ ±0.02 Å, CN$_{(Co-O)}$ ±20%, R$_{(2nd\,shell)}$ ±0.02 Å, CN$_{(2nd\,shell)}$ ±20%
%Res: deviation between experimental data and fit given by the relative residual in percent.
N= number of data points, Y$_{exp}$ and Y$_{theo}$: experimental and theoretical data points, respectively.

$$\% \text{Res} = \frac{\sum_{i=1}^{N}|y_{exp}(i) - y_{theo}(i)|}{\sum_{i=1}^{N} y_{exp}(i)} *100$$

Fe-K edge measurements

Bulk-XANES spectra at the Fe K-edge were collected at the Dutch Belgium Beamline (DUBBLE) at the European Synchrotron Radiation Facility (ESRF) in Grenoble, France. The measurements were collected at room temperature in transmission (ionization chambers) and in fluorescence mode (9 channel monolithic Ge-solid state detector).

µ-XANES data were collected on beamline 10.3.2 at the Advanced Light Source (ALS), Berkeley, USA (6). The measurements were collected at room temperature in fluorescence mode using a 7-element Ge-solid state detector with a beam size of ~5x5 µm^2.

All beamlines are equipped with a Si(111) crystal monochromator. The monochromator angle was calibrated by assigning the energy of 7112 eV to the first inflection point of the K-absorption edge spectrum of Fe metal foil.

Figure S6.2 shows reference compounds for Fe(0), Fe(II) and Fe(III). Bulk-XANES data at the Fe K-edge were collected on both Co(II)-doped and non-doped HCP samples. The data reveal that in both cement matrices Fe(III) is the predominant Fe oxidation state present. Figure S6.2 also shows Fe K-edge µ-XANES collected at Co(II)-rich spots (e.g., Co spot 1), at Co(III)-rich spots (e.g., Co spot 2) and at Co-poor spots (e.g., Co spot 3, not shown on the µ-XRF map). Again, Fe is present solely as Fe(III) in the Co-enriched-spots of the cement matrix.

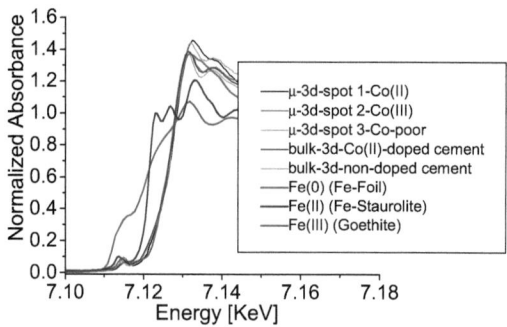

Figure S6.2. Fe K-edge bulk experimental spectra of reference compound for Fe(0), Fe(II), Fe(III) together with bulk-XANES of 3 days (3d) hydrated Co-doped and non-doped HCP and µ-XANES data of single spots rich in Co(II) and Co(III) as well as a single spot poor in Co, but rich in Mn.

Electron Paramagnetic Resonance (EPR)

The X band cw EPR measurements were conducted at the Physical Chemistry Laboratory, ETHZ, using an ElexSys spectrometer (Bruker, microwave frequency 9.453 GHz) with use of a liquid helium cryostat (Oxford). The spectra were collected at 77 K with a microwave power of 0.2 mW, a modulation frequency of 100 kHz and modulation amplitude of 0.1 mT.

The EPR technique is sensible to Mn(II) oxidation state. Mn(II) shows six major peaks in the magnetic field between 300 to 380 mT (7), marked with arrows in Figure S6.3. Figure S6.3 shows EPR spectra of Co(II)-doped HCP samples hydrated from 1 hour up to 150 days. The spectra clearly reveal six peaks, indicating that Mn(II) is present in the matrix. The intensity of these six peaks decreases with ongoing hydration time. Note that the intensity of the peaks is an approximate indication of the amount of, e.g. Mn(II), present in the matrix. Thus, the EPR study suggests that the relative amount of Mn(II) in the cement sample decreases as the hydration time progresses.

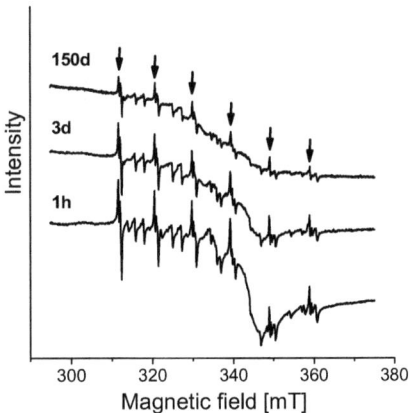

Figure S6.3. Electron paramagnetic resonance data of Co-doped HCP samples prepared in air and hydrated from 1 hour up to 150 days. The arrows indicate the six typical intensity peaks for Mn(II).

Co(II)-sorbed onto HCP

The hydrated cement powder (HCP) employed for the experiment was stored in the glovebox under nitrogen atmosphere. The Co(II)-sorbed onto HCP sample was also prepared in the glovebox. The sample was prepared by adding a Co(NO$_3$)$_2$ solution (0.3 mol/L) to the hydrated cement powder resulting in a Co concentration of 5000 ppm. The sample was then shaken end-over-end for one day. Note that the water used for the solution was not purged with nitrogen as for the case of the Co(II)-tot-O$_2$ sample.

Bulk-EXAFS measurements at the Co K-edge were collected at the Swiss Norwegian Beam Line (SNBL) at the European Synchrotron Radiation Facility (ESRF) in Grenoble, France. The measurements were collected at room temperature in fluorescence mode (Lytle detector). The beamline is equipped with a Si(111) crystal monochromator. The monochromator angle was calibrated by assigning the energy of 7709 eV to the first inflection point of the K-absorption edge spectrum of Co metal foil.

Figure S6.4 shows a selected k^3-weighted, normalized, background-subtracted bulk-EXAFS spectra of CoOOH reference compound and Co(II) sorbed onto hydrated cement reacted for 1 day. The experimental spectrum shows strong similarities to Co(III)OOH, indicating that the sorbed Co(II) oxidized to Co(III).

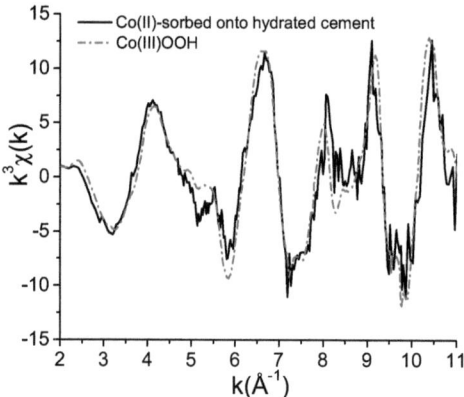

Figure S6.4. Co K-edge k^3-weighted, normalized, background-subtracted EXAFS spectra of Co(II)-sorbed onto fully hydrated cement and reacted for 1 day (in black) and CoOOH as reference (in red).

S6 References

(1) Baes, C. F.; Mesmer, R. E. *The hydrolysis of cations*; John Wiley & Sons, Inc.: New York, **1976**.

(2) Hem, J. D.; Roberson, C. E.; Lind, C. J. Thermodynamic stability of CoOOH and its coprecipitation with manganese. *Geochimica et Cosmochimica Acta* **1985**, *49*, 801-810.

(3) Manceau, A.; Schlegel, M.; Nagy, K. L.; Charlet, L. Evidence for the formation of trioctahedral clay upon sorption of Co^{2+} on quartz. *Journal of Colloid and Interface Science* **1999**, *220*, 181-197.

(4) Manceau, A.; Drits, V.; Silvester, E.; Bartoli, C.; Lanson, B. Structural mechanism of Co^{2+} oxidation by the phyllomanganate buserite. *American Mineralogist* **1997**, *82*, 1150-1175.

(5) Manceau, A.; Llorca, S.; Galas, G. Crystal chemistry of cobalt and nickel in lithiophorite and asbolane from New Caledonia. *Geochimica et Cosmochimica Acta* **1987**, *51*, 105-113.

(6) Marcus, M.; MacDowell, A. A.; Celestre, R.; Manceau, A.; Miller, T.; Padmore, H. A.; Sublett, R. E. Beamline 10.3.2 at ALS: A hard- X-ray microprobe for environmental and material sciences. *Journal of Synchrotron Radiation* **2004**, *11*, 239-247.

(7) Ottaviani, M. F.; Montalti, F.; Romanelli, M.; Turro, N. J.; Tomalia, D. A. Characterization of starburst dendrimers by EPR. 4. Mn(II) as a probe of interphase properties. *Journal of Physical Chemistry* **1996**, *100*, 11033-11042.

CHAPTER 7

CONCLUSIONS

A multi-analytical approach (microscopic, macro- and micro-spectroscopic) was used in this thesis to determine the immobilization process of Ni and Co in the complex cement matrix over a wide range of reaction conditions. The major findings and consequential conclusions are summarized below.

Microscopic investigations showed that in the Ni-doped cement matrix newly-formed Ni phases are always enriched around the clinker mineral alite and its hydrate product inner-C-S-H, regardless of the reaction conditions. This finding suggests that the clinker phase alite and the hydrated phase inner-C-S-H play an important role in the formation of the newly-Ni-formed phases. A tentative explanation is given by the specific features of the mineral alite (Ca_3SiO_5).

    i)   Alite dissolves readily upon contact with water, thus acting as a highly reactive zone in the cement matrix. This reactive zone is expected to have a high specific surface area. Therefore, a large number of surface sites may be exposed for Ni binding, which may facilitate Ni accumulation at the grain boundaries.

    ii)   It is speculated that, in addition to Ni, Al may accumulate in this zone due to rapid dissolution of Al-containing clinker phases, which could act as the Al source for the formation of the Ni-Al LDH.

The macro and micro-spectroscopic investigations revealed that for all reaction conditions only minor amounts of Ni-hydroxide formed, whereas Ni-Al LDH was the major component in the cement matrix. The only exception was observed in the Ni-doped cement with low Ni-loading. The latter, showed that at low Ni concentrated regions an unknown Ni species formed. Nevertheless, at highly enriched Ni-regions within the overall low concentrated sample Ni-Al LDH was still observed. Moreover, the amount of Ni-Al LDH was found to increase with ongoing hydration, indicating that Ni-Al LDH is the thermodynamically most stable form in the cement matrix, controlling the solubility of Ni in the system. The enhanced stability of Ni-Al LDH versus Ni-hydroxides in highly alkaline systems such as cement is attributed to the substitution of Al for Ni in the octahedral layers.

These findings are important for waste management performance assessment and are of major environmental relevance. The formation of a thermodynamically stable mineral phase, which controls the solubility of Ni, should be, therefore, taken into account in developing strategies for waste management. Thus, the formation of such a mixed Ni-containing solid phase, eventually as solid solution with varying composition, can significantly reduce the bioavailability and mobility of the heavy metal into the environmental systems.

The study on the Co speciation in Co(II)-doped cement performed under oxidizing conditions showed that Co(II) oxidizes to Co(III). After only 1 hour of cement hydration about a quarter of the added Co inventory was oxidized, indicating that Co(II) oxidation in cementitious systems is a fast process. With ongoing cement hydration the amount of Co(III) phases was found to increase. After 1 year more than half of the added Co inventory was present as Co(III). This finding indicates that in highly alkaline systems such as cement and in an oxidizing environment Co(III)-containing phases preferentially form. The geochemical implication of this result is that the mobility of Co contaminants in the environment in alkaline and oxidizing conditions will be strongly retarded, due to the decreased solubility of Co(III) compared to Co(II) phases. The finding of the rapid Co(II) oxidization in cementitious systems further indicates that for the storage of Co-containing waste packages produced under oxidizing conditions and stored in contact with air both Co(II)- and Co(III)-containing phases should be considered if short-term aspects (less than a few hundreds of years) have to be addressed in waste management performance assessment studies.

Additional experiments showed that the Co(II) oxidation can only be avoided if special measures are taken to achieve an oxygen free atmosphere during sample preparation. These experiments showed the necessity of careful checks of the oxidation state of redox sensitive elements in conjunction with uptake studies and or investigations of sorption processes by using appropriate spectroscopic techniques, e.g. X-ray absorption spectroscopy.

Micro-spectroscopic investigations were carried out to determine the Co(II) and Co(III) precipitates. The experiments indicate that Co(II) is incorporated in Co-hydroxide and/or Co-phyllosilicate phases, whereas Co(III) precipitates as CoOOH and/or Co-phyllomanganate. These findings contradict those resulting for the Ni-doped cement system, where the formation of Ni-Al LDH was observed. This indicates that Ni and Co, although both divalent metal cations with similar atomic numbers, exhibit a chemically different behaviour in an alkaline environment such as cement. It is therefore a priori not possible to assume that the immobilisation processes of all divalent metals will be similar in cementitious materials. Instead, this study clearly shows that it is essential to investigate each metal independently. Only a combination of wet chemistry methods with advanced spectroscopic and microscopic methods will provide the sufficiently detailed information to develop a fundamental mechanistic understanding of metal uptake processes.

OUTLOOK

The present study showed that a combined analytical approach is essential to obtain a better understanding of the immobilization processes of heavy metals in the cement matrix. In particular, this thesis further contributes to the knowledge of the uptake behaviour of Ni and Co in cement on the molecular level. Nevertheless, some questions raised during this study remained open.

The study demonstrated that the formation of Ni-Al LDH within a cement matrix is particularly related to the dissolution of the clinker mineral alite. A similar observation was made for Co-containing phases within the Co(II)-doped cement matrix in preliminary microscopic investigations. This finding indicates that the dissolution of alite during cement hydration plays an important role, independently of the heavy metals phases forming in the matrix. It has to be noted that alite is one of the major clinker phases. It dissolves readily upon contact with water forming the hydrated product of inner-C-S-H, which grows as rim around alite. Similar rims were observed for the metals uptake process. Thus, it appears that alite acts as a highly reactive zone in the cement matrix. This reactive zone is expected to have a high specific surface area. Therefore, a large number of surface sites may be exposed for the binding of metals and other elements, such as Si or Ca, facilitating the accumulation of these elements at the grain boundaries. It would be, therefore, important to further study metal uptake by the mineral alite, its chemical, mineralogical and kinetic properties, and sorption mechanism. Uptake by alite could be an important step and the driving force for all heavy metal or radionuclide uptake mechanisms in the cement matrix.

The study further revealed that at lower total Ni concentration an unknown Ni species formed in addition to Ni-Al LDH and small amounts of Ni-hydroxides. It is important to check whether this species corresponds to a Ni surface complex on a cement mineral, i.e. adsorption of Ni on any of the available cement phases. For the identification of this Ni species further experiments are needed. For example, experiments on the uptake of Ni on specific cement minerals, such as outer-C-S-H, at concentrations below the Ni solubility limit in the system could provide more insight into the structural properties of the Ni species.

Furthermore, this experiment at low Ni loading demonstrated that it is important for investigations on the metal uptake by heterogeneous matrices to carry out experiments also on dilute samples. If such experiments would not be carried out, some processes could remain concealed.

This study revealed the complexity of the chemical behaviour of Co(II) uptake by cement. Due to this complexity the exact Co speciation forming in the cement matrix could not be identified:

i) Co(II) oxidized to Co(III) within hours of hydration, thus two different oxidation states were present in the cement matrix and were detectable by macro and micro-spectroscopic methods (bulk- and µ-XAS).

ii) No pure Co(II) phase nor Co(III) phase was detected, but rather a mixture of phases for both oxidation states was observed, predominantly Co(II)-hydroxide and/or Co(II)-phyllosilicate and Co(III)OOH and/or Co(III)-phyllomanganate.

The question remains: is it possible to exactly identify the Co(II) and Co(III) species forming in the hydrating cement matrix? And if Co(III)-phyllomanganates are forming, what is the role of Mn in the Co-uptake process? Note that from the decrease of Mn(II), observed from the EPR data, it cannot be concluded that the formation of Co-phyllomanganates is not likely in the Co(II)-doped cement matrix. This can be attributed to the fact that Co(III)-phyllomanganates can either contain a single Mn oxidation state (Co-asbolane: Mn(IV)), or occur as minerals with mixed Mn oxidation states (e.g., Co-buserite: III and IV; Co-lithiophiorite: II and IV; Co-birnessite: II,III and IV).

To answer these questions further experiments on simplified systems could help to clarify the problematic. For example, experiments could be performed were measures are taken to obtain the presence of only one Co oxidation state per cement sample. Of course, such experiments should take advantage of the information gained by µ-XAS and µ-XRF. Furthermore, if at µ-XAS/-XRF beamlines both KB mirrors and zone plates could be implemented, complementary information could be gained. The use of KB mirrors allows collecting information on the overall distribution of metals in heterogeneous matrices by applying µ-XRF, whereas µ-XAS allows obtaining the full wealth of information on the speciation (coordination numbers, type of backscattering neighbours, bond distances, system disorder) on the micro-scale.

By exchanging the KB mirrors system with zone plates, elemental distribution maps from the same sample and the same investigated area with nanometer-scale resolution could be collected. This would, furthermore, allow identifying the different species and their oxidation state in heterogeneous matrices using the fingerprint method with XANES. With the improvement in spatial resolution (from 1x1 µm² to <50x50 nm²), therefore, it could be expected to detect single species.

The investigation of the metal uptake by cement by µ-XRD on thin sections revealed that at the present the µ-XRD data analysis is only partly achieved. This is partly due to the fact that, for example, the species identified during this thesis are in the nanometer range. The X-ray microprobes (µ-XAS/XRF) available at the 3rd generation synchrotron facilities have currently a beam size in the micrometer size range. Consequently, several particles are likely to be exposed simultaneously to the beam. If zone plates would be implemented at beamlines equipped for µ-XRD data collection, beam sizes down to 20 nm will become achievable. Thus, allowing the exposure of single particles to the beam.

A further problem in the µ-XRD data analysis is the arbitrarily oriented crystals probed from the X-ray beam. The random orientation of the crystals makes it impossible to compare the experimental data with the available diffraction tools and database. A software package is needed which includes peak searching, autoindexation used to find the right orientation matrix, integration, profile refinement and crystallographic database searching. The development of such a novel tool in the future could allow the complete extraction of long-range information obtained from the high resolution µ-XRD images.

MIX
Papier aus verantwortungsvollen Quellen
Paper from responsible sources
FSC® C105338

Printed by Books on Demand GmbH, Norderstedt / Germany